T0235911

Science, Technology and Medicine in Modern History

General Editor: John V. Pickstone, Centre for the History of Science, Technology and Medicine, University of Manchester, England (www.man.ac.uk/CHSTM)

One purpose of historical writing is to illuminate the present. At the start of the third millennium, science, technology and medicine are enormously important, yet their development is little studied.

The reasons for this failure are as obvious as they are regrettable. Education in many countries, not least in Britain, draws deep divisions between the sciences and the humanities. Men and women who have been trained in science have too often been trained away from history, or from any sustained reflection on how societies work. Those educated in historical or social studies have usually learned so little of science that they remain thereafter suspicious, overawed or both.

Such a diagnosis is by no means novel. Nor is it particularly original to suggest that good historical studies of science may be peculiarly important for understanding our present. Indeed this series could be seen as extending research undertaken over the last half-century. But much of that work has treated science, technology and medicine separately; this series aims to draw them together, partly because the three activities have become ever more intertwined. This breadth of focus and the stress on the relationships between knowledge and practice are particularly appropriate in a series which will concentrate on modern history and on industrial societies. Furthermore, while much of the existing historical scholarship is on American topics, this series aims to be international, encouraging studies on European material. The intention is to present science, technology and medicine as aspects of modern culture, analysing their economic, social and political aspects, but not neglecting the expert content which tends to distance them from other aspects of history. The books will investigate the uses and consequences of technical knowledge, and how it was shaped within particular economic, social and political structures.

Such analyses should contribute to discussions of present dilemmas and to assessments of policy. 'Science' no longer appears to us as a triumphant agent of enlightenment, breaking the shackles of tradition, enabling command over nature. But neither is it to be seen as merely oppressive and dangerous. Judgement requires information and careful analysis, just as intelligent policy-making requires a community of discourse between men and women trained in technical specialities and those who are not.

This series is intended to supply analysis and to stimulate debate. Opinions will vary between authors; we claim only that the books are based on searching historical study of topics which are important, not least because they cut across conventional academic boundaries. They should appeal not just to historians, nor just to scientists, engineers and doctors, but to all who share the view that science, technology and medicine are far too important to be left out of history.

Titles include:

Julie Anderson, Francis Neary and John V. Pickstone
SURGEONS, MANUFACTURERS AND PATIENTS
A Transatlantic History of Total Hip Replacement

Roberta E. Bivins
ACUPUNCTURE, EXPERTISE AND CROSS-CULTURAL MEDICINE

Linda Bryder
WOMEN'S BODIES AND MEDICAL SCIENCE
An Inquiry into Cervical Cancer

Roger Cooter
SURGERY AND SOCIETY IN PEACE AND WAR
Orthopaedics and the Organization of Modern Medicine, 1880–1948

Science, Technology and Medicine in Modern History
Series Standing Order ISBN 978–0–333–71492–8 hardcover
Series Standing Order ISBN 978–0–333–80340–0 paperback
(*outside North America only*)

You can receive future titles in this series as they are published by placing a standing order. Please contact your bookseller or, in case of difficulty, write to us at the address below with your name and address, the title of the series and one of the ISBNs quoted above.

Customer Services Department, Macmillan Distribution Ltd, Houndmills, Basingstoke, Hampshire RG21 6XS, England

The Politics of Addiction

Medical Conflict and Drug Dependence in England since the 1960s

Sarah G. Mars

*Qualitative Project Director, Heroin Price and Purity Outcomes Study,
University of San Francisco, California, USA, and Honorary Research Fellow, London
School of Hygiene and Tropical Medicine, UK*

First published 2012 by
PALGRAVE MACMILLAN

Palgrave Macmillan in the UK is an imprint of Macmillan Publishers Limited, registered in England, company number 785998, of Houndmills, Basingstoke, Hampshire RG21 6XS.

Palgrave Macmillan in the US is a division of St Martin's Press LLC, 175 Fifth Avenue, New York, NY 10010.

Palgrave Macmillan is the global academic imprint of the above companies and has companies and representatives throughout the world.

Palgrave® and Macmillan® are registered trademarks in the United States, the United Kingdom, Europe and other countries.

ISBN 978-1-349-30688-6   ISBN 978-1-137-27221-8 (eBook)
DOI 10.1057/9781137272218

This book is printed on paper suitable for recycling and made from fully managed and sustained forest sources. Logging, pulping and manufacturing processes are expected to conform to the environmental regulations of the country of origin.

A catalogue record for this book is available from the British Library.

A catalog record for this book is available from the Library of Congress.

10   9   8   7   6   5   4   3   2   1
21 20 19 18 17 16 15 14 13 12

For Jason

Contents

Figure and Tables

Figure

Tables

Abbreviations

ACMD	Advisory Council on the Misuse of Drugs
AIDA	Association of Independent Doctors in Addiction
AIP	Association of Independent Prescribers, later renamed the Association of Independent Practitioners in the Treatment of Substance Misuse
CIO	Chemist inspecting officer
CFI	Central Funding Initiative
CURB	Campaign on the Use and Restriction of Barbiturates
DDU	Drug Dependency Unit (colloquially known as a 'Clinic')
DH	Department of Health (the Department of Health split from the Department for Social Security in 1988).
DHSS	Department of Health and Social Security (formed by a merger of the Ministry of Health with the Ministry of Social Security in November 1968)
DTTO	Drug treatment and testing order
GMC	General Medical Council
GP	General Practitioner (NHS in this context)
Inspectorate	Home Office Drugs Inspectorate
KCWHA	Kensington and Chelsea and Westminster Health Authority
LCG	London Consultants Group
PCC	Professional Conduct Committee (of the General Medical Council)
RMO	Regional Medical Officer
SCODA	Standing Conference on Drug Abuse
script	Prescription
T&R	*Treatment and Rehabilitation* report of the Advisory Council on the Misuse of Drugs, London: HMSO, 1982.
Tribunal	Home Office Misuse of Drugs Tribunal

Glossary

Addiction/dependence Addiction and dependence are terms that were subject to considerable controversy in the 20th century but they are used interchangeably in this book to mean that the user of a drug is diagnosed by a doctor as having adapted to its presence and would suffer if it were withdrawn abruptly.

Clinic This is a colloquial name for the hospital-based Drug Dependency Units (DDUs) set up in the late 1960s and early 1970s.

Consultant (medical) In this context 'consultant' refers to the highest grade of hospital doctor in England. Despite the name, consultants are salaried employees, usually in charge of a hospital department or unit.

Controlled drugs These are drugs controlled under the Misuse of Drugs Act, 1971. These include heroin, cocaine, methadone, amphetamines and benzodiazepines. The term replaced 'dangerous drugs', which was used in 20th-century domestic legislation until 1971.

Drug In this context a drug is a psychoactive substance used in either an illegal (according to the Misuse of Drugs Act, 1971) or an unsanctioned way. This includes heroin, cocaine, solvents and tranquillisers but for convenience excludes alcohol and nicotine.

Drug doctors This term is used to denote any doctor with significant involvement in treating drug-related problems. NHS psychiatrists are sometimes referred to as 'addiction specialists' although not all the psychoactive drugs used by their patients are addictive and some of the patients are not treated for addiction but for other drug-related problems. The term does not imply the prescribing of substitute drugs, although this may be involved.

Methadone Synthetic opiate, also known as Physeptone, used to prevent withdrawal symptoms in opiate addicts. It is most commonly prescribed as an oral liquid but it also comes in an injectable form and as oral tablets.

Opioid This term covers both derivatives of the opium poppy such as morphine and heroin ('opiates'), and pharmacologically similar synthetic substances such as methadone.

Private prescriber This is a doctor paid by fee outside the NHS who prescribes substitute drugs (opiates, stimulants and tranquillisers) to patients for the treatment of addiction. They may be a general practitioner or have specialist training in addiction psychiatry. They may work concurrently in the NHS.

Substitute prescribing This is usually used to describe the prescribing of one drug to replace another, such as methadone for heroin. However, here it also describes prescribing the same drug, such as heroin, as it is often intended to replace or obviate the need for a trafficked supply of the drug.

Trafficked drugs This is a term used in the 1980s and in this book to differentiate between pharmaceutically produced substances obtained legally or illegally by users, and drugs smuggled or 'trafficked' into Britain from producer countries.

Acknowledgements

This project was only made possible with intellectual, material, technical and emotional support that has spanned two continents. In Britain, I would like to thank Virginia Berridge, my PhD supervisor, whose wise advice, stimulating discussions and painstaking approach enabled the navigation of many pitfalls and made this research a rewarding and enjoyable experience. The Economic and Social Research Council generously funded the research project from 2000 to 2003 through grant number R000239228. Thanks are also due to my colleagues at the London School of Hygiene and Tropical Medicine – Stuart Anderson, Jennifer Stanton, Nicki Thorogood and Ornella Moscucci – as well as members of the London School of Hygiene and Tropical Medicine's History Group, whose knowledge and insight have been valuable. I am also grateful to Ingrid James, Jennifer Richards and Janet Gardner for their assistance in overcoming all manner of administrative obstacles.

In the drugs field I would like to thank many colleagues and friends, especially John Witton, Alan Quirk and Jim McCambridge, for their practical help and constructive discussions; I am also especially grateful to John Strang for his early and impartial encouragement of the project. Essential to the success of this research were the staff of the Department of Health's Archive in Nelson, Lancashire, particularly Pauline Connor, and at the General Medical Council; library staff at DrugScope (formerly the Institute for the Study of Drug Dependence); members of the Department of Health Clinical Guidelines Working Groups for their loan of committee papers, many of which were missing from government archives; staff at the Home Office, in particular John Corkery and Alan Macfarlane; Detective Sergeant Geoffrey Monaghan; James Tighe and Matthew Johnson for their help in making contact with private prescribers; and to Dale Beckett, the late Tom Waller and the late Arthur Banks for the loans of historical materials. I am greatly indebted to the many interviewees who kindly agreed to talk on and off the record and at length about their recollections.

In San Francisco, my deep gratitude goes to Dorothy Porter, who acted as an informal PhD advisor while I was away from London, for her kind support, encouragement and helpful advice during my work's transformation into a published book. Brian Dolan gave generous and invaluable advice on securing publication. I am also grateful to Judith Barker who kindly helped host my move to the University of California, San Francisco (UCSF). Library staff at UCSF were patient and tolerant of my extensive demands on their time and I would like to thank Andy Panado in particular. Thanks also are due to Jenny McCall, Clare Mence, Michael Strang and Ruth Ireland at Palgrave Macmillan and to John Pickstone for his helpful comments on the manuscript.

Last, and most of all, I owe the greatest debt to my family: to my parents, Valerie and Gerald, for their unstinting support of my education, formal and informal, and for their sensitivity throughout this research; to my parents-in-law, Nicole and Jeffrey, for all they have done to help; to my brother, Adrian, for his invaluable, timely technical support and for generously providing the beautiful cover photograph; and, of course, to my husband, Jason, and daughter, Mia, for their encouragement and forbearance.

Introduction

If you are sick, who decides what treatment you should receive? Who controls where treatment should be given and by whom? Do you as a patient have a choice, or is it all in the hands of doctors, civil servants or even politicians? If you were seeking help for drug dependence, would the same rules apply? This book examines how decisions about addiction treatment have been made in England. In particular, it asks how the conflicts that arose between a few doctors towards the end of the last century affected national addiction treatment policies.

The use of opiates and other currently controlled substances was not always considered a matter for medical attention, or even a significant problem for their users. In the 18th century, opiates could be bought over the counter from ordinary grocers' shops and were used for the relief of a range of ailments. Many forces have shaped certain patterns of drug use into what we recognise as 'addiction', and made doctors into the appropriate source of assistance for that problem.[1]

Since the 19th century, England's medical profession has played a major role in this area, prescribing drugs such as heroin and morphine to addicts considered unable to give up using them. In 1926 a committee of doctors appointed by the government considered whether long-term prescribing not aimed at eventual abstinence constituted medical treatment. It concluded that drug addiction was an illness rather than a criminal activity and recommended that the prescription of substitute heroin or morphine to addicts in non-increasing doses should be allowed to continue for those patients who needed it. The report's authors considered that for some addicts, 'attempted complete withdrawal produced severe distress or even risk of life', while others became incapable of work.[2] This practice of prescribing opiates to addicts and its loose regulatory framework became known as the 'British System'.

1

While it may not have constituted a 'system' in terms of a centralised policy or set of rules, the clinical freedom it allowed doctors and the underlying spirit of compassion towards addicts certainly existed in the minds of many of those in policy and treatment. Although it is often poorly defined, few would consider that treatment of addicts today still constitutes the British System. Debates also took place in the USA regarding whether addiction was an illness and the use of maintenance doses but with the opposite result. The US Supreme Court outlawed maintenance prescribing in 1919 and doctors were only allowed to supervise short-term withdrawal until many decades later.[3]

An addict seeking medical help in early 1970s England might have visited his general practitioner (GP), who would have sent him either to a private doctor or to one of the new specialist NHS Drug Dependency Units (DDU), known as 'the Clinics'. These were housed in hospitals under the leadership of psychiatrists and originally prescribed injectable heroin and methadone, along with barbiturates and amphetamines.[4] The Clinics had been set up with the aim of undercutting the black market by prescribing the drugs that addicts were using. Until the mid-1960s the illicit market was mainly fed by the overflow from doctors' prescriptions. It was not until the early 1970s that an established market in illicitly imported heroin developed.[5]

Liberal prescribing was also to be found among private prescribers, although they were rarely allowed to prescribe heroin or cocaine. However, by the late 1970s and early 1980s, an opiate addict would have had quite a different reception at the NHS Clinics. New patients were offered only decreasing doses of oral methadone aimed at detoxification and abstinence, often with a compulsory requirement to attend counselling. Meanwhile, private doctors were continuing to offer addicts long-term prescriptions for oral and injectable drugs. Tension arose between private and NHS doctors over what constituted the most appropriate treatment.

By the late 1990s, an addict would find yet another range of options. An NHS GP might be willing to prescribe methadone on a long-term basis, often with the help of a local Clinic. Some GPs might not want to provide any treatment, referring an addict to a Clinic or private prescriber instead. The Clinics had changed their practice too, offering long-term opiate prescribing that sometimes included heroin, and even the experimental prescribing of amphetamines to stimulant addicts.[6]

In the first years of the 21st century, patients of private prescribers might have found themselves seeking alternative care after their doctor was disciplined by the General Medical Council. Between 2000 and

2010, 6 of the 11 practising private prescribers with significant patient numbers had been struck off the medical register. In 2004 the Council's largest disciplinary case ever charged seven private prescribers with serious professional misconduct after the death of a patient.[7] By 2010, little private prescribing remained in England, and patients might have found treatment from the expanding non-statutory or voluntary sector.

What lay behind these changes? Was the increase in drug use or the spread of HIV/AIDS influential? Did research evidence guide change? Why were so many private prescribers disciplined? This book examines how the process of regulating the medical profession was used to shape policies and practice, for it was through regulation by the state and the profession itself that battles were fought over the proper roles of NHS and private prescribers.

For many years, NHS doctors accused private practitioners of prescribing substitute drugs in excessive quantities, with the risk of causing overdose in their patients and excess drugs being sold on – spreading addiction to others. These doctors were also portrayed as entering the field without adequate training or experience, of failing to check their patients' compliance with treatment and of being motivated by money.[8] Because most dependent drug users in treatment were unemployed, it was also argued that they must be selling some of their prescribed drugs on the black market in order to pay their medical bills. In turn, private doctors accused the NHS of hypocrisy, of being overly bureaucratic and of caring more about controlling the supply of prescribed drugs than about the health of their patients.[9] Hearings before the General Medical Council and the Home Office's Drugs Tribunals, the medical press and the general media all took a turn in airing these arguments.[10,11] Media coverage in turn fed into the regulatory process.

Terms used

In the published literature, the debate between doctors working privately and those working for the NHS in the drugs field has been portrayed as a clash of sectors – 'public' against 'private' – particularly when seen from the viewpoint of consultant psychiatrists working exclusively in the NHS. Official documents have also distinguished private prescribers from NHS doctors as individuals. However, the interviews carried out for *The Politics of Addiction* have shown that many of the doctors involved in the private sector also worked in the NHS, and some had lengthy careers in the NHS before their private practice. Of the private doctors prescribing to drug users and paid by fee interviewed between 2000 and 2003, more than half (8/14) had been, or were still, working for the NHS (see Appendix).

I avoid the word 'sector' not only because of this overlap between public and private but also because the private 'sector' was in fact wider than those doctors involved in the 'public-private' debate. Private psychiatric hospitals provided abstinence-oriented detoxification but, since this rarely involved the prescribing of substitute pharmaceuticals (such as methadone), they could not be held responsible for supplying the illicit market in prescribed drugs, or for overdose deaths, and they have remained uncontroversial, outside the debate. Blurring these boundaries further, in the 1990s, private doctors also received patient referrals from the NHS, and occasionally might be paid by social service departments or from other public funds to carry out their work. In the 21st century the voluntary sector has increasingly employed doctors to prescribe. This book therefore uses 'private prescriber' to denote those doctors whose treatment involved prescribing substitute drugs for addiction and who accepted fees for their services, whether specialists or generalists, and 'drug doctor' for any medical professional significantly involved in treating drug problems with, or without, prescribed drugs. 'GP' refers to a general practitioner working in the NHS, while the private GP is included in the term 'private prescriber'.

A note on geography

Private prescribing has been virtually unknown outside London and the South East of England. Although much private healthcare has traditionally been focused in the metropolis, it was surprising to find such a stark contrast between the South East and the rest of the country. A market explanation seems unconvincing as there would have been plenty of demand from the large populations of drug users in cities outside the South East from the 1980s onwards. Even if demand were low, private practice could be undertaken by NHS GPs on a small scale alongside their usual practice. Despite extensive enquiries, no conclusive answer to this puzzle has been found, but it might be explained by differences in enforcement by the Home Office Inspectorate and the police. Both organisations had regional structures, and some interviewees have suggested that tolerance of private prescribing was greater among Home Office inspectors in the South East than outside. Yet, although private prescribing was a metropolitan phenomenon, it was usually discussed as a national issue with national policy implications.

Historical questions

Where did the conflict between NHS and private doctors originate? What lay behind the major policy changes of the 1980s and what part did opposition to private prescribing play in their formation? How did medical self-regulation function in the highly polarised field of addiction treatment? How did the state regulate the 'self-regulating' medical profession? What can we learn about doctors' strategies and values from their less formal groupings, and how did these affect their failures and successes in these battles? How and why did regulation change in the last years of the 20th century in the treatment of addiction, and why is there virtually no private prescribing left in the 21st century? What can these developments tell us about conflict, regulation and power? *The Politics of Addiction* addresses these questions.

1
1965–2010: A Background Sketch

Introduction

From the 1930s through the 1950s, while England faced and fought fascism in Europe and rebuilt in its aftermath, the country's drug scene was relatively peaceful. The rest of the 20th century, however, saw a dramatic transformation in the ways drugs were obtained and used, stimulating growing public and professional interest. Between 1970 and 1999 there was a massive increase in the availability of illicit drugs in England,[1] and a corresponding rise in the numbers of drug users both outside and seeking treatment.[2] From the 1980s there was a new disease that could be transmitted through injecting drug use: AIDS. These three developments were central to the policy changes during this period.

In 1970, 2657 addicts were notified to the Home Office, while in 1992, after a long rise, 24,703 addicts were notified. Estimates vary as to whether these figures represent half, one-fifth or one-tenth of the overall addict population.[3] The impact of HIV/AIDS, once its transmission through injecting drug use became clear, was major. HIV was known to have infected drug users in New York by 1984–85 and a few deaths had occurred in Britain. In 1985, concern significantly permeated the drugs 'policy community'[4] – the network of civil servants and experts involved in making and advising on policy in and around government. The reality of HIV's arrival became clear when an epidemic among injecting drug users in Edinburgh was made public in 1986. Complex political manoeuvring preceded the official permission for syringe provision to drug users and the subsequent allocation of specific funding to HIV prevention.

A fragile national consensus emerged which emphasised a pre-existing and more accepting approach to drug use, while attempting to reduce the harm it caused to the user and others, becoming known as 'harm minimisation' or 'harm reduction'. Prescribing was used to attract patients into treatment services, with the Department of Health promoting a return to the prescription of oral methadone on a maintenance basis to discourage injecting at a time when long-term prescribing was discouraged. Needle exchanges, which had sprung up through grassroots activism, were introduced officially, albeit on a 'pilot' basis.[5] The drugs field, long divided between those advocating abstinence as the goal of treatment and those more sympathetic towards maintenance prescribing, saw a shift towards greater consensus after HIV and in 1988 the harm-reduction approach received official policy approval.[6]

1965–1970: The second Brain Committee

The committee responsible for a new age in drug treatment services, the Interdepartmental Committee on Drug Addiction, was chaired by Lord Brain, a former president of the Royal College of Physicians (1950–57). It published its slim report to government, the second Brain Report, in 1965.[7–11] Its membership and the almost wholesale implementation of its recommendations by government reflected the dominance of the medical profession in the formulation of drug treatment policy in the first three-quarters of the 20th century. In 1961 the same committee's first report had advocated a medical rather than a criminal justice approach to drug users, recommending treatment in the psychiatric ward of a general hospital because 'addiction should be regarded as an expression of mental disorder rather than a form of criminal behaviour'.[12] This medical approach was reinforced in the 1965 report with its statement that 'the addict should be regarded as a sick person [and] should be treated as such and not as a criminal, provided that he does not resort to criminal acts'.[13]

As a response to the growing number of drug users of a noticeably different social and age demographic, the committee reconvened in 1964. Since the 1920s there had been very little opiate addiction. What there was had tended to be concentrated among 'therapeutic addicts' who had acquired their dependence inadvertently through medical treatment, and among professionals involved in medicine whose proximity to medicines had facilitated their dependence. They were a diminishing, ageing population who received prescribed drugs and were not generally seen as a cause of social disturbance. Fears were raised by the press and parliament, however, in response to the new type of young, usually male, drug users, mainly congregating in London from the late 1950s. Between 1960 and 1964, the number of heroin addicts known to the Home Office rose from 94 to 342.[14] The number of cocaine addicts also increased from 52 in 1960 to 211 in 1964.[15] Today, these figures seem startlingly tiny but the increases warned of things to come.

The 1965 Brain Report, reconsidering its earlier findings, resulted in wide ranging legislative and policy changes. The committee's medical membership interpreted its terms of reference 'as meaning that we were not being invited to survey the subject of drug addiction as a whole, but rather to pay particular attention to the part played by medical practitioners in the supply of these drugs'.[16] The report concluded that the major source of the new addicts' heroin and cocaine was not trafficked drugs but 'the activity of a very few doctors who have prescribed excessively for addicts'.[17] Greater treatment provision and tighter control of supply within a medical framework were the report's recommendations, implemented in the 1967 Dangerous Drugs Act and the Dangerous Drugs (Supply to Addicts) Regulations, 1968, which introduced special licences, to be granted by the Home Office to doctors wishing to prescribe heroin or cocaine.

Between 1968 and 1970, specialist hospital-based drug dependency units (DDUs) were set up, mostly led by consultant psychiatrists and generally in London where the problem was particularly concentrated. In practice, the Home Office almost exclusively limited heroin and cocaine licences to doctors working in the DDUs or the 'Clinics', as they became known, and in hospital departments. Until this point, many addicts were known by the Home Office through doctors' voluntary reports, inspections of pharmacy registers and inspectors' face-to-face contacts with users. From 1968, formal notification became a statutory requirement modelled on infectious disease notification.

The problem of drug use was defined as that of addiction, maintaining the disease model. The second Brain Report described addiction as 'a socially infectious condition'. It has been argued that prior to the 1960s the medical model was only pursued in terms of individual treatment but that the second Brain Report formulated the disease model to emphasise control within a public health approach.[18] These developments drew drug users into specialist medical treatment and discouraged GPs from involvement, moves not forcefully opposed by the latter.[19] They also established the DDUs in a dual role of treating drug users and controlling the wider drugs supply to addicts. This control system saw the Clinics as near monopoly suppliers of drugs,[20] which not only prescribed free drugs but also checked a government register to see whether a patient was already receiving a supply from another doctor.[21]

1970–1984

The Misuse of Drugs Act, 1971, was a substantial piece of legislation, consolidating previous Dangerous Drugs Acts, renaming them 'controlled drugs' and incorporating heavy criminal penalties. It created an important policy mechanism in the Advisory Council on the Misuse of Drugs (ACMD), taking over from the earlier Advisory Committee on Drug Dependence established in 1967,[22] to advise on future policy responses to the evolving drug scene. It brought back the Home Office's Drug Tribunals, which were designed to regulate doctors' prescribing of controlled drugs behind closed doors.[23]

In the early years of the DDUs, the numbers of addicts were very small with only 2240 registered heroin addicts in 1968, and the Clinics seemed able to meet patient demand.[24] They were initially liberal in their prescribing, but policy began to change in the mid-1970s when they started offering only short-term detoxification to new patients, while private doctors picked up patients' unmet demand for long-term prescribing.[25] Major changes also took place in England's illicit drug supply: until 1979, prescribing had remained the main source of opiates and other drugs both legitimately and on the illicit market, with patients selling or sharing the excess from their prescriptions. H. B. 'Bing' Spear, Chief Inspector at the Home Office between 1977 and 1986, recalled that some expensive smuggled Chinese heroin could be found but relatively small quantities of trafficked drugs were entering the country.[26] However, from 1978 to 1979, the quantity of trafficked heroin in England increased,[27] as did the number of heroin users both outside and seeking treatment.

Until the 1980s, most of England's heroin use and its treatment provision had been concentrated in London, but where heroin spread across the country, drug services were slow to follow.[28] What Clinics there were had insufficient treatment places and found that drug users were increasingly looking elsewhere for treatment. The Home Office Addicts Index[29] showed that during the 1970s the proportion of patients seeing both private and NHS GPs grew in both absolute terms and as a proportion of all those seen by doctors. After the establishment of the Clinics, NHS doctors in general practice had had little involvement in the treatment of addiction and minimal training. In 1970, GPs only notified 15 per cent (111) of all addicted patients to the Addicts Index in 1970. This rose to 29 per cent (264) of notifications in 1975 and 53 per cent (1191) in 1981.[30]

Outside the NHS

While government and the medical profession chiefly shaped services within the NHS, the voluntary and private sectors tended to play the role of meeting unmet demand. The term 'voluntary sector' has been used here to encompass charities and other non-statutory, non-profit organisations. Voluntary bodies set up to help drug users with social and health problems were numerous in the drugs field. The late 1960s saw a growth in street services and day centres providing social care and counselling in London and other cities, some church based, usually following a social rather than medical model but often with close links to treatment services.

The UK's first Narcotics Anonymous (NA) began in 1979, modelled on Alcoholics Anonymous, a '12-step' or 'Minnesota Model' fellowship. These meetings aimed at maintaining daily abstinence from all mood-altering drugs, with attendance and 'recovery' going long beyond initial detoxification. Psychiatrist Brian Wells, a 12-stepper himself, described a common cynicism both among users and professionals regarding NA in the early 1980s. Despite this the movement continued to grow.[31] Voluntary services were represented by the umbrella organisation, the Standing Conference on Drug Abuse (SCODA), set up in 1973.

Those working within the NHS were also involved in voluntary sector projects and their approaches had mutual influence. Griffith Edwards, an NHS psychiatrist who had started and run the Institute of Psychiatry's Addiction Research Unit, was instrumental in establishing Phoenix House, an abstinence-based therapeutic community modelled on its original in New York. Fellow NHS psychiatrist John Strang has suggested that these and other similar abstinence rehabilitation houses in the UK influenced the move against maintenance prescribing in the late 1970s and early 1980s.[32]

A system under strain

Despite the expansion of specialist care from London to the provinces (by 1975 there were 15 outpatient DDUs in London and 21 in the rest of the country), the continuing increase in the number of drug users put pressure on their ability to meet demand,[33] and Clinic staff felt a sense of therapeutic disillusionment. In the country at large, optimistic expectations about future investment in the health service had been ended by the impact of the 1973 oil crisis on the British economy. Historian Charles Webster explained, 'Until that time, it was confidently anticipated that the economic system was capable of achieving a rate of growth sufficient to meet rising social expectations.'[34] Optimism did not return swiftly as from 1974 to 1979 four factors created a state of crisis and demoralisation in the health service: cuts in public expenditure; Sir Keith Joseph's reorganisation; resentment from vulnerable groups about the failure to improve services; and the failure in leadership of health ministers.[35] From 1974 the hospital service lost its protection from spending cuts, and health spending plans became subject to stricter financial disciplines.[36]

The second Brain Report had anticipated that controls on the prescription of heroin and cocaine would be sufficient to deal with demand, but, once the prescription of those drugs was under tighter control, there seems to have been a move among patients to obtain other drugs from doctors outside the Clinics. The Iranian Revolution with its resultant emigration helped to establish a new heroin route into Britain from the Gulf, meeting demand of existing addicts who were no longer supplied by the DDUs, and spreading use across the country on a previously unimagined scale. This source was then superseded by Turkish heroin in 1980 and then the following year's major supplier became Pakistan.[37]

Yet it is perhaps unsurprising that a medical committee which had restricted its remit to the role of medical practitioners in the supply of drugs, rather than 'drug addiction as a whole',[38] did not consider or anticipate the subsequent changes in the international drugs trade. As the DDUs had been set up with the aim not only of treating but of controlling the spread of addiction, the penetration of trafficked heroin into new areas of the country in the 1970s, and most dramatically from 1979, provided a basis for the criticism that the Clinics had failed. In some circles, this was presented as a failure of the 'medical model'.[39] Others responded by criticising maintenance prescribing, about which they had long felt uncomfortable.[40]

The reluctant return of general practice

In 1968, GPs had lost the authority to prescribe heroin and cocaine to their addict patients, although they could still prescribe them for the treatment of pain and some other indications. Other opiates, like methadone, could be prescribed by any doctor for the treatment of drug dependence. Until the 1980s, most general practices in England and Wales had had little to do with the management of drug misuse. The opposite was the case in Scotland, where there was minimal specialist involvement.[41] Due to the relatively small numbers of drug users in the

1970s, few GPs in England were affected by the problem, but by the early 1980s the situation had changed and heroin addicts sought help from their GPs, bringing them into the picture in an unplanned way.[42]

The ACMD addressed this state of affairs in 1982 and recommended that renewed GP involvement become official policy alongside the Clinics.[43] The government responded to these recommendations and an ongoing battle began between forces encouraging GP involvement (emanating from both specialists and generalists, the drug policy community and central government) and the many reluctant GPs, supported in the 1990s by their trade union, the General Medical Services Committee of the British Medical Association. GP reluctance was largely due to the unpopularity of drug addicts as patients and uncertainty over whether drug problems constituted an appropriate sphere for medical intervention, even among those who treated them as patients.[44] Similar attitudes have been noted in doctors' attitudes towards alcoholic patients, described in sociologist Philip Strong's study of doctors and 'dirty work'.[45]

Polydrug use and the Clinics

In the 1970s a pattern of use distinctive to Britain emerged, with drug users injecting barbiturates often in combination with other drugs. The hypnotic and tranquilliser drugs used became seen as a major problem for accident and emergency departments, particularly in London, due to frequent overdosing and aggression towards casualty staff.[46,47] Through the 1970s, barbiturates were the drugs most commonly involved in overdose deaths among addicts. After experimentation, it was concluded that barbiturates were not a suitable drug for maintenance therapy through the Clinics, which were later criticised for their apparent inability to respond to polydrug use and, in particular, barbiturate use.[48] Whether, in fact, polydrug use was a new phenomenon in the 1970s or had always been part of the non-therapeutic drug use addressed by the second Brain Committee was unclear. Gerry Stimson and Edna Oppenheimer noted that in 1964 virtually all the cocaine users known to the Home Office were also addicted to heroin.[49]

In 1975 the ACMD launched the Campaign on the Use and Restriction of Barbiturates (CURB) to reduce barbiturate prescribing by doctors. According to Bing Spear, 'As an effective response to the barbiturate-injecting problem, CURB was a singularly futile exercise, which merely postponed the day when realistic controls would have to be imposed.'[50] Barbiturates eventually became controlled drugs in 1984, but by this time the problem had already diminished, possibly because of the increasing availability of trafficked heroin in the 1980s.[51] As barbiturates fell from favour, benzodiazepines were mistaken for a non-addictive substitute[52] and, with widespread prescribing, their use by addicts followed suit. By 1986–87, benzodiazepines were commonly available from GPs and on the streets.[53] In Scotland in particular, a 'non-injectable' gel-filled oral temazepam capsule was formulated to prevent this use but persistent injectors suffered horrific injuries and disease during the 1980s and 1990s. In 1992 the ACMD called for restrictions on the prescription of temazepam but legislative change did not follow until three years later. An alternative and, in the eyes of the BMA, very effective approach to restricting the black market in temazepam gel-filled capsules was pursued by banning the formulation from NHS prescription.[54]

1982 onwards: political interest grows

Under the Conservative government, drugs became a highly political issue with substantial resources allocated to services, high-profile media campaigns and the first comprehensive government strategy document for drugs policy in 1985. Responding favourably to the recommendations of the Advisory Council on the Misuse of Drugs,[55] the Department of Health and Social Security (DHSS) prepared a large new source of funding to cover start-up costs for new services. This Central Funding Initiative (CFI) consisted of £17.5 million distributed in 188 grants from 1983 to 1989.[56] It aimed at funding local initiatives, such as the development of cross-agency problem drug teams, the development of community-based responses across the country, and integration of drug services into mainstream health services. On the quiet it was also intended to shift the concentration of services and power away from the London psychiatric Clinic consultants.[57]

The voluntary sector gained new status and recognition in the mid-1980s.[58] Although acknowledging the importance of the CFI, David Turner, who represented voluntary drug services from 1975 to 1994 as co-ordinator of SCODA, considered that the sector's strong influence and growth pre-dated the flow of money from the CFI by a couple of years.[59] However, it may be that he preferred to see voluntary services as responding sensitively to local demand rather than following central edict. This initiative and the return of GPs have also been linked to a 'normalisation' of drug services in first half of the 1980s, as drug use and drug dependence became more common and drug services were integrated into mainstream healthcare.[60,61]

1985–1999

British drug policy during the 1980s has received academic interest from sociologists, anthropologists and historians.[62–65] Agreement has emerged over a number of the themes of this period: that community drug services, both voluntary and statutory, expanded during the 1980s; that the professional groups involved in drug treatment and policy increased and diversified; that GPs re-entered the picture after more than a decade's absence; and that in response to HIV/AIDS, drug and treatment policies were liberalised in the late 1980s, with 'harm reduction' becoming official policy in 1988. Later observations by Gerry Stimson, a sociologist and activist in the harm-reduction movement, defined 1997 as the beginning of yet another new phase, with the election of Blair's New Labour government. This, he claimed, brought an end to the 'public health approach', dating from 1987, where 'the aim was to help problem drug users to lead healthier lives, and to limit the damage they might cause themselves or others', and introduced an 'unhealthy' 'punitive and coercive ethos' for dealing with dependent drug users.[66]

Behind these policies, drug use continued to rise, to spread to new parts of the country and diversify. New drugs and new formulations joined the existing array of substances, while others dropped from availability or favour. Heroin use climbed through the 1980s and 1990s, joined by 'crack', a new smokeable form of cocaine, which came from the USA in the mid-1980s and grew to considerable popularity. Ecstasy (the street name for 3,4-methylene-dioxymethamphetamine), a stimulant with empathy-inducing properties, became popular as a 'dance drug' at parties and clubs, usually taken as an oral tablet, along with other stimulants and psychedelic drugs. Amid great public and media concern over a small number of sudden deaths associated with the drug, educational responses were launched but no individual treatment was provided. Meanwhile, cannabis remained the most popular drug in England throughout this period, with demands for reduced penalties or legalisation becoming increasingly common and less controversial.

GPs and community-based services

From the beginning of their re-involvement, with the exception of a small number of enthusiasts, and despite concerns over HIV/AIDS in the later 1980s and 1990s, GPs remained reluctant to prescribe substitute drugs to addicts. In 1990, GP Tom Waller, prominent for his encouragement of his peers, proposed additional payments to GPs as an incentive for treating drug users.[67] Although criticised as expensive, possibly unethical and probably ineffective,[68] the idea was taken up by GP negotiators in 1996, who declared that treatment of drug misuse was no longer to be considered part of their contract to provide general medical services but required an additional fee.[69] While there were a few local arrangements paying extra, the Department of Health did not move on the issue.

Despite GPs' wariness of addicts, commentators noted a shift from specialist to generalist services during the 1980s. The 1984 clinical guidelines and subsequent DHSS circulars reinforced this, making drug services more like other areas of the NHS, where it was unusual for any condition to be addressed solely by specialists. Criminologist Alan Glanz has linked the revival of GPs' involvement in drugs and the emphasis on 'community', rather than specialist or institutional care, to their rising status as a group. GP leaders had been working to establish general practice as a 'speciality' with academic departments and compulsory vocational training. Improved terms and conditions had followed and by 1984 it had changed from being an unpopular career choice for medical students to the most desirable.[70]

Political encouragement of private medicine, which strengthened through the 1980s and 1990s, related mainly to services reimbursed by health insurance rather than direct payment by the patient and did not concern private prescribers. Early after achieving power, the Conservatives abolished the Health Services Board, established by Labour to supervise the private hospital sector and phase out private beds from the NHS,[71] but private prescribing was almost overwhelmingly on an outpatient basis.

AIDS and official harm reduction

Once those in the drugs field had started to see HIV/AIDS as an important threat, a number of policy options were available. Hard-line campaigns against drug use had issued from the Conservative government in 1985–86 and at the same time a penal approach, at both a political rhetorical and a policy level, pressed through legislation to freeze, trace and confiscate money from drug dealing and to increase penalties for trafficking.[72] Virginia Berridge's research has shown that while a continuation of this penal and stigmatising approach might have been expected from the New Right, in the event it was a non-coercive public health approach that won out. The struggle behind this owed much to medical bureaucrats in the Department of Health in alliance with outside pressure groups in the voluntary sector. As a result,

AIDS brought together politicians and 'experts' in an alliance based on minimising the harm from drug use, rather than eradicating or curing it, using needle exchange as the means to achieve this.[73] Although given the new name of 'harm reduction', this approach had a long history, with antecedents in the 1880s and 1960s.[74] Rather than the drugs policy community switching wholesale from one approach to another, controversy over different methods of dealing with drug use had existed since at least the 1960s, with different groups gaining ascendance at particular moments. 'Fixing rooms', for instance, where injectors could take their prescribed drugs, had existed in the early 1970s, but along with the provision of injecting equipment, these had been phased out by 1975 as the Clinics moved to providing oral drugs.[75] The voluntary sector had always pursued a more 'harm reductionist' approach but advocated it more openly after 1986.[76]

The significant policy event of that year was the McClelland Report, from a committee set up by John Mackay at the Scottish Home and Health Department under the chairmanship of Dr D. B. L. McClelland. From a committee membership not derived from the drugs field, it was this document which first officially championed a harm-reduction approach in relation to AIDS, including the establishment of needle exchanges. This position has often erroneously been given to the ACMD, whose report *AIDS and Drug Misuse* did not come out until 1988.[77,78] Scotland had taken the lead on this approach as the problem of HIV among injecting drug users had been effectively publicised by Dr Roy Robertson, a GP practising in the deprived Muirhouse area of Edinburgh. In 1985 he had found levels of HIV among his injecting patients of around 50 per cent.[79]

Harm reduction, which became official British policy in 1988, changed prescribing once again. AIDS made long-term prescribing a legitimate option once more and appeared to resolve 'the prescribing question that had bedevilled drug policy in the 1970s and 1980s'.[80] The 1960s and 1970s policy of 'competitive prescribing' was revived to attract drug users into treatment, albeit with oral methadone rather than injectable heroin. Just as proponents of harm reduction did not appear overnight in 1988, neither were its earlier opponents complete converts under the new 'consensus'. Furthermore, 'harm reduction' meant different things to different professional groups.[81] Political scientist Hervé Hudebine noted that the 1991 edition of the clinical guidelines,[82] chaired by John Strang, emphasised the importance of harm reduction but reasserted abstinence as a primary goal, and advised GPs against undertaking methadone maintenance without specialist advice. Through this the specialists, who had had to face competition from other sectors in both financial and policy terms since the first half of the 1980s, reaffirmed their primacy.[83]

Part of the government's strategy against HIV/AIDS involved funding research not just on epidemiology and biology but also on the intimate behaviour of drug users, including their injecting and sexual practices.[84] Government research grants went from a total of £2.5 million in 1986/7 to around £23 million in 1992/3.[85] By 1995–96, however, Hudebine noted that harm reduction, still pursued at local level, had almost disappeared from the national policy agenda and that earmarked funds for Health Authorities to prevent AIDS also ceased after 1993.[86] This was then followed by harm reduction becoming more contentious once again in the political rhetoric, and it had fallen to the lowest ranking policy goal of the White Paper *Tackling Drugs Together* by 1995.[87] Sociologist Nigel South has observed, however, that harm reduction continued as a policy priority in Scotland.[88]

Drugs and crime

While possession and distribution of drugs controlled under the Misuse of Drugs Act, 1971, were usually crimes in themselves,[89] public and policy concern over drug-related crime during this period tended to translate as acquisitive crime perpetrated to obtain the means to buy addictive drugs, and sometimes violent crime resulting from intoxication. Estimates varied as to what proportion of crime was committed by drug users in pursuit of their substances. In the mid-1990s, politicians and drugs policy researchers produced contradictory estimates, with researchers emphasising the range of income sources available to dependent heroin users other than acquisitive crime.[90,91] In the late 1990s, however, there seemed to be emerging consensus in the drug policy field, as well as among politicians, on the importance of links between dependent drug use and acquisitive crime. A literature review showed that dependent heroin users, disproportionately likely to be poor people in deprived communities, were very likely to resort to burglary, shoplifting, fraud and theft to pay for drugs.[92] Stimson observed with dismay the changes he observed in treatment services that flowed from making this connection. Focusing treatment on reducing drug use in order to curb drug-related crime broke the post-AIDS public health consensus, which had prioritised the prevention of bloodborne disease and pursued harm reduction as a humanitarian goal.[93]

While some of Stimson's concerns related to anticipation of the future direction of such policies, policy initiatives were already in place by the end of the century. Drug treatment and testing orders (DTTOs), influenced by American 'drug courts', could 'sentence' a drug user to treatment rather than prison, with freedom dependent on monitored results, and these were piloted by the Criminal Justice Act, 1998. Without waiting for the pilot study's conclusions, the Home Secretary extended DTTOs across the country. Until this point there had been little coercive treatment in England,[94] although it had been discussed since the 1880s and was recommended by the second Brain Committee. Another linking mechanism used in the 1990s was arrest referral schemes, where drugs workers sought out drug users in the criminal justice system, often in police cells, and referred them to treatment. Here, though, involvement with the schemes was voluntary and not an alternative to prosecution. Although these multiplied from 1999 onwards, they had been in existence before this, and some have seen arrest referral as part of a liberal rather than a penal approach.[95]

So has Stimson overemphasised the starkness of policy change from public health to crime prevention in the pre- and post-Blair era to make a political point?[96] Berridge took the view that penal policy persisted during the era of harm reduction, albeit in a modified form, and that coercive approaches to drug and alcohol treatment had their roots as far back as late nineteenth-century inebriates legislation. Between 1987 and 1997, Britain did not depart from the international or European systems of drug control and, at a local level, police were involved in drug advisory committees, co-operating in the establishment of needle exchanges.[97] Furthermore, the option of diverting drug users into treatment rather than prison had become government policy as long ago as 1990 in the White Paper *Crime, Justice and Protecting the Public*. Berridge, writing in the early 1990s, considered the balance of power between penal and medical approaches post-AIDS to be too complex to be 'adequately subsumed under rhetorical barriers, such as the "public health" approach of drug policy'.[98] Furthermore, Stimson overlooked the potentially coercive role of public health, which has used powers of compulsory quarantine and notification.

Voluntary services

Voluntary services became critical to the direction of policy and service provision post-HIV, although initially divided over the issue of needle exchanges.[99] The distinction between 'voluntary' and 'statutory' had become somewhat blurred in the last quarter of the 20th century by government funding of voluntary sector organisations. This trend strengthened in the 1980s when the Conservative government started to contract out many statutory services to the voluntary sector. SCODA's David Turner claimed that the establishment of voluntary services had not diminished their role as advocates of drug users and agitators for change. And, although government funding could be seen as a way of controlling these organisations, and reining in their radicalism, Berridge, in her work on the anti-tobacco pressure group Action on Smoking and Health, has shown how state support for a radical group outside government could serve to lobby for change desired by, but unvoiceable from, government.[100] Turner, writing after needle exchange had become orthodoxy, explained voluntary drug services' fears over endorsing harm reduction as a result of threats to funding when they were perceived 'as having gone too far',[101] suggesting that control was still an element in state funding.

Professionalisation was a feature of the 1980s, and continuing in the 1990s, in the voluntary sector, including greater requirement for formal qualifications among staff, management standards, performance measures and other bureaucratic features demanded by those contracting its services. Also emerging in the 1990s was drug user activism, agitating for changes to services and legislation.[102] As well as providing statutory services, the voluntary sector saw the growth of self-help groups in the 1980s and 1990s. NA continued to spread across the country with 223 weekly meetings by 1991. There were also residential 12-step treatment centres in the private and voluntary sectors, with 'a diluted version' sometimes found in NHS addiction units. By 1991 there were 30 treatment centres in the UK and Ireland providing Minnesota Model drug-free-style treatment.[103]

Local arrangements

In the 1990s, central government encouraged treatment services to make arrangements locally, and chief among these exhortations was 'shared care', which involved a formal division of a patient's workload between specialist psychiatrists and GPs.[104] Local inter-agency co-operation had been encouraged for many years but from 1995 there was a radical departure from the established arrangements, with the setting up of Drug Action Teams in every health district. Their memberships comprised a small number of budget holders ideally representing key local authorities, services and criminal justice agencies. Their aim was to reduce drug-related harm in accordance with the targets set by the Conservative government's White Paper *Tackling Drugs Together*. These goals were aimed at reducing both drug supplies and demand for drugs, and they encompassed both penal and harm-reduction approaches.

Each Drug Action Team was advised by a Drug Reference Group made up of local people with expertise in the various services, and these arrangements persisted through to the end of the century with minor modification. Similar but separate arrangements were set up following strategies for Wales, Scotland and Northern Ireland. Later, under Labour, Drug Action Teams became responsible for commissioning and evaluating drug services.

Wider changes in health services, public and private

If drug treatment services had joined the mainstream in the 1980s, what was happening in the rest of the health service? A major theme of the 1980s and 1990s in the rest of the NHS was the changing relationship between the centre and the periphery, with management becoming increasingly important. Before the 1974 reorganisation of the NHS, 'management was conspicuous by its absence'. Administrators and treasurers did not take a proactive line in developing services, which was left to the medical profession.[105] This was followed by a period of 'consensus management' that tended to reinforce the strong position of the medical profession, but all this changed with the election of the Conservative government in 1979. From then on the NHS underwent 'continuous revolution'.[106] The medical profession's assumed right to consultation over NHS changes was not honoured by Margaret Thatcher and even employment terms and conditions were imposed without mutual agreement.[107]

General management was introduced in 1984–85, providing for the first time, according to Stimson and Lart, an effective central mechanism for controlling peripheral activity beyond budgetary control. However, this central control paradoxically encouraged devolved decision-making, which in turn led to a huge increase in guidelines, directives and circulars from the centre, advising the periphery on how it was to carry out these devolved responsibilities.[108] The CFI could be seen as part of this pattern, encouraging the development of locally autonomous services while orchestrating them from the centre. Throughout the 1990s, management of the NHS was led by the NHS Executive, with centralisation becoming stronger in the second half of the decade.

Most controversial was the introduction of market reforms and a split between 'purchasers' of healthcare, GPs and Health Authorities, and providers, hospitals and community services, following 1989's White Paper *Working for Patients*. With providers' budgets dependent on the success of their services in attracting patients, the idea was that both consumer choice and efficiency would improve. From this major change arose a pressure to quantify the outcomes of treatment for comparison and to standardise treatment through the use of clinical guidelines, coinciding with the emerging 'evidence-based medicine' movement in the medical profession, which favoured guidelines as a distilled, applied source of research findings. The market endured under John Major's premiership but was partially dismantled by Tony Blair, reflecting its unpopularity with the public.

One of the themes of John Major's period of office noted by Rudolf Klein was the transformation of NHS patients into 'consumers'. The *Patient's Charter* (1991) outlined patients' consumer rights for the first time, although it was more symbolic and rhetorical in significance than in actually producing change. The extent to which NHS patients were able to exercise effective choice as consumers has been questioned.[109] Consumerism was also a popular theme with New Labour, appealing as it did across employees and employers, the constituents of 'old' Labour and the New Right.

With the rejection of competition as the spur of change in the NHS, the managerialism of the early and mid-1980s was revived in the late 1990s. The new National Institute for Clinical Excellence (NICE) was set up to assemble and disseminate evidence in good practice guidelines and policy advice.

Against this background of new and growing state controls over the medical profession, there came to light the case of two heart surgeons working at Bristol Infirmary. Found guilty of serious professional misconduct in 1997 after the deaths of 15 small children, the government capitalised on the case to increase scrutiny in the NHS without medical opposition. On top of the huge media attention, the government launched a public inquiry into the case, creating an atmosphere in which the medical profession were pushed into accepting a much higher degree of government control than ever before in the NHS. Clinical audit, where the outcomes of treatment were monitored, was made compulsory.[110] In 1999, trust in the profession was further shaken when GP Harold Shipman was accused of mass-murdering his patients over a long period.[111]

Although government attention fell directly on the public sector, the increased pressure on the General Medical Council (GMC) also increased surveillance of all doctors. By the end of the 20th century, medical regulation looked quite different from how it had 30 years earlier: the president of the GMC himself was calling for a more active approach to self-regulation and the medical Royal Colleges had accepted regular competence testing of consultants. Klein concluded: 'collegial control over the performance of doctors had largely been maintained but at the cost of sacrificing the autonomy of individual doctors'.

Wider drug policies

In 1985 the first comprehensive drug strategy, *Tackling Drug Misuse*, had been published by the Conservative government.[112] This new development signalled increased political interest, and Stimson has claimed that this act politicised drug strategy in a new way,[113] but when the subsequent Labour government published its ten-year drug strategy, *Tackling Drugs Together to Build a Better Britain*, it demonstrated continuity with the Conservatives' earlier *Tackling Drugs Together*,[114] and cross-party consensus. The appointment to the newly created post of 'Drug Czar' of the former chief constable of West Yorkshire, Keith Hellawell, was seen as part of the penal approach to drug policy dating from 1997.[115] However, his deputy, Mike Trace, had extensive experience in drug treatment services. The 1998 drugs strategy departed from its predecessors by concentrating policy on heroin and cocaine as the drugs causing the greatest harm, and by hailing health interventions as the most effective way of reducing offending behaviour over and above penal solutions. Hellawell put forward performance targets for the next decade – for instance, the reduction of the number of people under 25 using heroin and crack cocaine by a quarter within five years and by a half within ten years. Such targets drew criticisms from a number of sources as unmeasurable by existing mechanisms.[116] They were later quietly abandoned, as was, less quietly, the Drug Czar himself.

Those who have passed judgement on the 1990s have tended to emphasise continuity over change.[117,118] Perhaps because they have considered drug policy as a whole, rather than focusing on treatment services, any move away from harm-reduction rhetoric and greater use of coercion in treatment were marked as less significant than in the work of Stimson.[119] Though Nigel South acknowledged a punitive approach in both rhetoric and legislation, he saw inconsistency in policies across Britain. Labour's concerns about the role of 'social exclusion' as a factor in drug use were seen by both Rowdy Yates, a harm-reduction activist, and Geoffrey Pearson, a criminologist and sociologist, as a significant change during the late 1990s,[120,121] but what impact this had in practical policy terms was unclear. Both authors also considered the emergence of ecstasy and the widespread dance drug phenomenon of the late 1980s and 1990s as a major development, which Yates claimed had 'made existing drug treatment services almost irrelevant'.

How treatment policy was formulated, 1970–1999

The drug policy community and the policy-making process have been considered primarily by Stimson and Lart, Berridge, Smart, Duke and MacGregor.[122–126] Stimson and Lart noted the traditions of British policy-making which continued into the 1970s, reached through committees where debate was characterised by politeness and an absence of politics. Policy was made in private through accommodation between experts and civil servants, as exemplified by the ACMD, set up in 1971. Berridge's account of the development of AIDS policy during the 1980s, although involving much more media attention and a greater variety of outside groups, had similar components being privately formulated between bureaucrats and outside interests and experts.[127] While doctors were not the chief architects of policy, as they were with the second Brain Report, key members of the profession, particularly medical civil servants like Dorothy Black, and psychiatrists like John Strang, held great influence. The growth of new drug agencies following the Central Funding Initiative drew many new occupational groups into working with drug users, diversifying the policy community in the 1980s, and displacing the purely medical perspective on drug use and users.[128] Responses to drugs in the late 1980s included a more prominent place for government, the criminal justice system, and the community, with medicine taking an important but less central role.[129]

In a departure from the earlier 'gentlemanly' period of policy-making, Stimson saw the late 1980s as a time of politicisation. The establishment of the Ministerial Group on the Misuse of Drugs, for instance, showed that drugs were moving out of professional and advisory committees and that debate was becoming more public.[130] Linked to this politicisation was a huge rift between the 'political' and 'policy' community view of drugs, exemplified by the controversy over the Conservative government's mass media anti-heroin campaign in 1985–86. Going against 'expert' advice from the drugs policy field, including that of the ACMD, which opposed widespread publicity not part of an overall educational approach,[131] the advertisement told people that 'Heroin Screws You Up', with the aim of eradicating rather than reducing the harm from use. The government commissioned its own evaluation of the campaign, which gave it positive results, but the methodology was also criticised by the policy community.[132] Undeterred, in 1987 the government launched another campaign with the message 'Don't Inject AIDS'. These events corresponded with anthropologist Susanne MacGregor's picture of a

British approach to policy developing from debate among a limited range of 'well-informed interest groups' which shared a basic consensus. This process would occasionally be interrupted by intervention from politicians seeking to gain political capital from taking up drug issues.[133]

Examining both national policies and local drug services in London in the last 15 years of the century, Hudebine described the policy process as existing at a number of levels simultaneously, with gaps between the levels of national political rhetoric, policy resulting from civil servants and from local agencies. A complex process appeared to be at work in the drug policy community, involving various understandings, tolerance and flexibility, and acceptable degrees of confrontation and challenge born of mutual dependence between government and the various agencies. This allowed some degree of coexistence within the apparent policy contradictions of the different levels.[134]

2000–2010

The first decade of the 21st century did not see the astonishing rises in drug use witnessed in the 1980s or 1990s. Drug use levelled out or fell among 16–59-year-olds asked about their consumption during the previous year (with the exception of crack and powder cocaine).[135] Services for drug users continued to evolve, with some particularly noteworthy trends. The voluntary sector grew to make up 35 per cent of the total costs of treatment, often involving collaborative arrangements with the private sector and the NHS. The Labour government's aim of shifting drug treatment into the voluntary and community sector resulted in the proportion of NHS treatment places falling from 80 per cent in 2001 to 65 per cent in 2005.

In the 20th century, prescribing policy had largely been left to medical leadership. However, in 2001, Labour politicians started to express views about the value of injectable heroin and other opiates for the treatment of addiction, which were then included in the 2008 government Drug Strategy.

Overall, drug policy continued to focus particularly on users of opiates and crack, which were designated the most problematic drugs.[136]

Conclusion

This background sketch of the last five decades has shown a period of turbulent change in both drug use and the policy responses to it. An increasing number and widening range of people have become involved in taking illicit drugs, in commenting upon drug use and in providing services. The policy process has moved from being conducted mainly in private to an often public and more overtly political undertaking, and while there was no disagreement about the ubiquity of drugs in the early 21st century, the extent to which their use has become 'normal' remains contentious.

2
Prescribing and Proscribing: The *Treatment and Rehabilitation* Report

No one had the faintest idea of what they were doing and were all expected to solve the problem of drug dependence.[1]

Dr Thomas Bewley

Introduction

It might be surprising that a new system of specialist, state-funded Clinics could be set up with so little idea of how to approach their task, but seen through the eyes of senior psychiatrist Thomas Bewley, such was the situation in the early days of the Clinics. According to the minutes of a meeting of Clinic leaders and civil servants in 1969, when discussing the possibility of changing the law to allow compulsory treatment of patients, 'the view was expressed that the philosophy and aims of treatment were at present too ill-defined for a decision to be reached on the subject'.[2] It was in this state of uncertainty that the Clinics tried a number of approaches such as cocaine prescribing which was quickly abandoned.[3] Heroin and methadone were prescribed in injectable form on a long-term maintenance basis and treatment could involve cocktails of stimulants and depressants, bargained over by doctors and patients.[4] With the Clinics prescribing generously to addicts without the many restrictions that were later introduced, their approach was closer to that of the private prescribers. Patients had less to gain from 'going private' and from 1968 to the mid-1970s there was a degree of peaceful co-existence between the Clinics and private prescribers. Criticism of other doctors by the Clinics, if expressed, tended to focus on GPs instead. At a Department of Health and Social Security (DHSS) meeting of Clinic leaders and civil servants, 'the prescribing of methadone to addicts by general practitioners was unanimously condemned', but private prescribers were not mentioned.[5,6]

When the Ministry of Health had advised doctors to prescribe heroin to addicts in order to prevent the large scale development of an illicit market in smuggled heroin in 1967,[7] most illicitly traded substances were either stolen or surplus from legitimate medical supplies. Prescribers could still influence the supply of drugs considerably because only a trickle of trafficked drugs was entering the country. So at the outset treatment in the Clinics had two aims: to help the individual patient and to protect wider public health by inhibiting the growth of a criminal black market. A leading Clinic psychiatrist involved in setting up the new system explained the individual approach to patients, saying: 'regular contact between the addict and the doctor of the centre gives the opportunity for a relationship to build up which may eventually lead to the addict requesting to be taken off the drug'.[8]

Optimism that addicts in treatment would eventually decide to give up using drugs may have been misplaced. By 1975 the Department of Health and Social Security observed that 'A pool of addicts on long-term maintenance who are unwilling to try to break their dependence on drugs has built up in the years since the present system was introduced in 1968.'[9] Clinic staff sought a fresh approach and in the later half of the 1970s, even before the new wave of heroin addiction, they started to take a more confrontational approach to their patients. At the same time they abandoned their attempts to contain the market in trafficked heroin.

The Clinics began to favour methadone over heroin, and oral, rather than injectable, formulations. Instead of maintenance prescribing, these services instigated a limited stabilisation period on a fixed dose that was then progressively cut to zero, often with a contractual obligation to attend for therapy.[10] Soon new opiate addicts entering London Clinics were offered only oral methadone detoxification without the option of longer term prescriptions or injectable drugs. (See Chapter 3 for more details of these prescribing changes.) As treatment allegiances solidified in the mid-1970s, the era of mutual tolerance between private prescribers and the Clinics drew to an end. From the late 1970s to 1982 lines of allegiance hardened.

Towards the end of the 1970s, smugglers started bringing large quantities of heroin into the UK, attracting new devotees to the drug. With the change in the main source of illicit opiates from doctors' prescriptions to trafficked drugs, doctors found that instead of being the chief guardians of the drug supply, they now faced major competition from a fully fledged black market in imported heroin, and a growing pool of demand across the country, exceeding the Clinics' ability to provide. What the Clinics were offering was also growing less attractive to addicts. Drug users were increasingly looking elsewhere for treatment. Following a decade or more of exclusion from treating addiction and without additional training, GPs were facing patients asking for treatment. Over the 1970s the proportion of patients seeing GPs practicing privately and in the NHS grew in both absolute terms and as a proportion of all those seen by doctors. Private doctors were also responding to the increase in demand, offering long-term maintenance prescribing of both injectable and oral methadone. Bing Spear dated disquiet over these perceived incursions into the Clinics' territory to 1979 when they were discussed at a London Consultants Group meeting.[11]

While treatment had become more uniform in the Clinics, doctors outside, both NHS and private, did not conform so easily. Initially in the medical press a vehement debate developed around the differences in prescribing methods of these groups of doctors. In 1980 the first open attack on private prescribing appeared in the *British Medical Journal*, in which its author, Thomas Bewley commented,

> There are strong economic pressures on addicts to try to obtain controlled drugs[12] on prescription and then to sell some of them; and there are subtle pressures on a doctor who considers prescribing privately to convince him that he will be treating patients rather than selling drugs ... The medical profession should consider whether there is any place for private treatment of addicts where a fee is contingent of a prescription.[13]

From his position in charge of St Thomas and Tooting Bec Hospitals' Drug Dependency Units (DDUs), Bewley recommended a list of safeguards to doctors, with special precautions for and about private prescribers. In order to stop 'script doctors', as they were derogatorily termed, he suggested restricting all psychoactive prescribing to 'licensed practitioners only', of whom he was one.[14,15] This would have effectively stopped such prescribing outside the Clinics.

The debate was revisited in *The Lancet* in January 1982, which said, of private doctors prescribing opioids, 'Their rationalisation is that the patient is thereby "saved" from the black market; however, since most addicts can only finance their private consultation by selling parts of their prescription, knowingly or (with a stretch of the imagination) unknowingly the doctor is prescribing sufficient drugs for this purpose.'[16] In the course of the debate private prescribers were accused of selling drug prescriptions for profit rather than treating patients.[17] Meanwhile private doctors accused the Clinics of failing to meet addicts' needs, being too rigid to allow individual doctors to prescribe to addicts where they thought it necessary. They claimed that Clinic doctors prescribed drugs in inappropriate formulations leading their patients into criminal activities to pay for black market supplies.[18] Later that year, *Treatment and Rehabilitation*[19] (known as *T&R*) proposed radical changes in services for drug addiction and new controls over prescribing doctors working outside the Clinics.

Treatment and Rehabilitation

The report came from the Advisory Council on the Misuse of Drugs (ACMD),[20] and attracted powerful ministerial support. At the DHSS, Secretary of State Norman Fowler's enthusiasm brought funding for many of its proposals.[21] The changes affected doctors across the country. A key motive among those crafting the new policies inside the ACMD was the growing dispute between the small number of private prescribers and NHS Clinic psychiatrists in London.

T&R expanded treatment services beyond the hospitals and back into the community after the centralisation of the late 1960s. It also outlined a role for voluntary services within a multi-disciplinary response, praising their 'problem-oriented approach' in contrast to the substance-based approach of the Clinics.[22] Both of these organisational changes had been suggested by the Labour government back 1975 but failed to attract funding to support them.[23] The policy-making process seen in the *T&R* Working Group and with the Clinical guidelines which followed centred around the 'expert committee'. This maintained the drugs field pattern of the 1960s and early 1970s, where decisions were reached through committees in private by accommodation between experts and civil servants.[24] Published research evidence played a role,[25] but the main emphasis was on the authority assumed integrity and non-partisan approach of the committee members.

T&R emerged in 1982 after seven years in the making. Its secretariat was mainly provided by the Home Office although it was published under the name of the DHSS and officials from both departments had attended the meetings.[26]

The *T&R* Working Group's completed project was preceded in 1977 by its more cautious preliminary findings.[27,28] This 'interim report' proposed retention of the existing system pending reviews and further research. It also laid the ground for some of the recommendations taken further in the final report, including its view that 'a multi-disciplinary approach to the problem of drug misuse is essential'.

The interim report did not discuss the form treatment itself should take, and avoided tackling the sensitive issue of substitute prescribing, saying 'We recognise that there is considerable uncertainty about effective methods of treatment for drug misusers and we avoid making specific recommendations which might seem to limit innovation.'[29] Government solicited comments on the interim report from Health Authorities, social services authorities and professional and voluntary organisations. David Turner, as founder member and representative of the SCODA representing voluntary sector drug services, was asked to identify areas for consideration in the final report taking into account their responses. Most of SCODA's member organisations were not medical and did not prescribe to drug users.

The resulting paper, signed off by David Turner, made some radical proposals against a background of Home Office statistics and responses to the report that apparently confirmed the interim report's view of a 'serious and slowly worsening problem'.[30] Turner drew attention to the need for information about the situation outside London and marked as a 'major dilemma' the Clinics' varying prescribing policies, with particular contrasts between those within London and those outside. He came to the radical conclusion that, 'The role of the treatment service ... as both a treatment system and as a means of control both of the supply of drugs to dependent persons and of the spread of addiction is no longer viable, if it ever was.'[31] The paper concluded by suggesting two alternatives to the *T&R* Working Group: adapting the present structure to make services available to a wider group of patients, or proposing an alternative model for the provision of services 'which is not based upon the substance misused but the social, medical, personal, etc. problems facing the individual'.[32] Given the tone of the paper and the preceding justification, much greater weight went behind the second of the two options. In a century where committees of doctors had shaped the major changes in British drug policy, a non-medical actor was questioning the value of the dominant medical system and laying the ground for a new phase in drug treatment policy.

Preparing the final report was an expanded *T&R* Working Group with a wider geographical spread ready for its new remit: 'to examine the range of services available for those who suffered harm through their drug misuse; consider whether this was sufficiently flexible to the needs of the individual and suggest ways in which the combined response could be improved'.[33] This second *T&R* Working Group had four psychiatrists, two members from social services, two nurses, a worker from the Citizen's Advice Bureau (married to a prominent social scientist with an hereditary title), a lecturer in social work, a psychologist, a GP, a probation officer, the director of a rehabilitation facility, a regional medical officer, a professor of oral medicine, the chairman of the ACMD and the co-ordinator of SCODA. Most of the second Working Group were already members of the ACMD.[34] The committee included no private practitioners but they provided oral evidence. While the mix of 'experts' and concerned, well-connected citizens on the *T&R* Working Group typified earlier policy-making styles in the drugs field, its multi-disciplinary membership was a departure from the all-medical Brain Committees of the 1960s.

Treatment and Rehabilitation: Its findings and significance

A number of 'significant changes' had taken place in the drug-using landscape since the 1960s. This led the *T&R* Working Group to question the second Brain Committee's model of 'treatment and containment'; in particular, multiple drug use, barbiturate and other tranquilliser misuse, the increase in the proportion of new heroin addicts in the numbers being notified to the Home Office Addicts Index[35] and a fall in the age of drug users. The increase in the proportion of addicts being notified to the Index from outside the Clinics prompted discussion of why drug users might be turning away from the DDUs and towards private practice or NHS GPs. It speculated that, particularly where there was no prospect of an addict becoming abstinent, curbs on prescribing by the Clinics might have encouraged drug users to seek treatment elsewhere in order to obtain prescriptions. It observed that 'many Clinics fall far short of the above minimum standards'.[36]

The solution to the Clinics' shortcomings was to be a reversal of the existing policy that had excluded GPs. General practice, already established throughout the country, offered a cheaper solution to extensive development of the Clinic system, although some hospital-based expansion was also recommended. To the *T&R* Working Group, the involvement of GPs offered wider geographical coverage and treatment for more drug users. However this also risked devolving prescribing decision-making away from the centre, justifying the report's measures to strengthen prescribing regulation. Yet these control measures were less aimed at future developments in general practice as at the existing situation in 1982: private doctors' perceived over-liberal prescribing and the black market in pharmaceutical drugs.[37]

Chapter 7 of the report proposed extensive curbs on prescribing by 'doctors working away from the hospital-based specialist services', that is to say NHS GPs and private prescribers. The chapter was most particularly concerned by 'a marked increase in private prescribing to problem drug takers, particularly in London, exemplified by three doctors in private practice who contributed over 10 per cent of all notifications to the Home Office during the nine months January to September 1980'.[38] The rise in treatment outside the Clinics worried the *T&R* Working Group for four reasons: a 'Lack of specialised knowledge, training and experience' essential for working in 'this difficult area';[39] the dispensing of drugs less often than daily increasing the likelihood of supplies being diverted to the black market and of drug users overdosing; pressure from patients on vulnerable doctors to prescribe drugs was listed as a worry, with uncited 'evidence of doctors issuing prescriptions simply to get rid of threatening patients' and finally the lack of 'easy access to the support staff' and facilities that were available to doctors in some hospital-based Clinics.[40]

The report saw the move from the Clinics as both supply- and demand-led. While partly blaming the Clinics for their limited prescribing, it also claimed that liberal prescribing was attracting patients away from the Clinics to obtain larger doses of drugs from other doctors. This, in turn, could increase their dependence and was, according to *T&R*, increasing the amount of legally manufactured drugs available in the illegal market as patients sold their surplus. Although it stated that 'problems arise whether the doctor provides treatment under the National Health Service or privately', *T&R* went on to vehemently attack private prescribing, even questioning whether a therapeutic relationship could develop when fees were involved.[41] It found existing regulatory mechanisms inadequate, remarking that private prescribing of controlled drugs to problem drug takers was 'undesirable' because there were 'moral and ethical aspects which cannot easily be dealt with by the General Medical Council (GMC) and give grave cause for concern'. There was suspicion as to how mostly unemployed patients could pay for treatment without selling a proportion of their prescribed drugs on the black market, although no evidence was cited.[42]

The report proposed three corrective measures: the preparation of 'good practice' prescribing guidelines by a Medical Working Group; the extension of Home Office licensing from heroin and cocaine to all opioid drugs, with urgent action being taken on dipipanone,[43] and changes to the Home Office Tribunal system so that it addressed a wider range of 'irresponsible prescribing'. This last recommendation may have been suggested by the Home Office Drugs Inspectorate,[44] representatives of which were present at the Working Group's meetings. Bing Spear later expressed his agreement with the report's criticism that the Home Office had underused the Tribunal system.[45] Home Office tribunals are discussed in detail in Chapter 5 of this book.

Since its publication in 1982, *T&R* has been defined as important in a number of ways. Its advocacy of integrating treatment and rehabilitation services through a multi-disciplinary approach involving health, social service, probation, education services, and the voluntary sector was widely seen as a departure from existing policy[46] but this was not a new idea. The second Brain Committee had recommended that long-term rehabilitation be integrated into the Clinics from the outset[47] but it was never done. Calls were again made in the DHSS's 1975 White Paper *Better Services for the Mentally Ill*[48] and by the *T&R* Working Group's interim report[49] but the split between treatment and rehabilitation remained through the rest of the century.

Both *T&R* and the interim report showed many areas of continuity with *Better Services for the Mentally Ill*, and indeed many documents written in the 1970s: the apparent increase in multiple drug use unmatched by services, overburdened Clinics and the ongoing drug use by long-term users despite treatment. Both the interim report, the 1975 White Paper, and later *T&R*, advocated a 'multi-disciplinary' approach within Clinics and between Clinics and other agencies; as did most of the policy documents that succeeded them, emphasising a wider approach to addiction beyond the medical into social rehabilitation. Also echoing *Better Services for the Mentally Ill*, there was a perception that drug services required central funding because in times of spending cut-backs, unpopular patient groups would be the first to suffer at the local level. This was repeated in *T&R*, winning the support of Norman Fowler in what became the Central Funding Initiative (see Chapter 1).

Commentators have given prominence to *T&R*'s redefinition of the 'drug addict' as the 'problem drug taker'. Following a change of terminology in the alcohol field, it described problem drug takers as 'any person who experiences social, psychological, physical or legal problems related to intoxication and/or regular excessive consumption and/or dependence as a consequence of his own use of drugs or other chemical substances (excluding alcohol and tobacco)'.[50,51] The Advisory Committee on Alcoholism had produced a report on the pattern and range of services for problem drinkers which was received by the *T&R* Working Group, in which the term 'alcoholic' was replaced with 'problem drinker'. Dr Anthony Thorley, one of the group's Clinic psychiatrists, was impressed and considered its equivalent might usefully replace 'addict' as a non-medical term.[52,53] It has been claimed that this eased the movement towards a more problem-oriented approach and away from a preoccupation with the particular substance being used.[54]

In conceptual terms, introducing the new term 'problem drug taker' seemed to recast the policy focus away from a disease-based model to a broader viewpoint. A less narrowly medical model might seem to limit the role of medicine by inviting input from the other professions and voluntary services. However, historian Betsy Thom has suggested that in the alcohol field this change also opened up new approaches for psychiatry,[55] and it seems that a very similar effect could be seen in the drugs field, with psychiatry maintaining a dominant, if challenged, position.

The 'problem drug taker' label could be seen as both normalising and re-pathologising drug users: on the one hand it suggested that not all drug users experienced problems with their drug use as 'the majority are relatively stable individuals who have more in common with the general population than with any essentially pathological sub-group'.[56] On the other it implied that addiction was not the only problems that drug services, both medical and non-medical, might need to address, adding regular excessive consumption and intoxication. The *T&R* Working Group's minutes showed that Dr Thorley was keen to put drug use into a wider context outside medicine,[57] and the report reflected this, arguing against the utility of the disease model: 'The problem drug taker seeking treatment may regard himself as having a disease or illness and may adopt a relatively passive sick role', which was 'inappropriate in the management of drug problems where clearly there is a volitional element, and personal responsibility and accountability are implicit.'[58]

T&R was probably credited with innovations that earlier policy documents had pioneered because, unlike its forerunners, this time many of them were implemented. Dipipanone was swiftly added to the list of drugs for which doctors needed a Home Office licence to prescribe in the treatment of addiction; a Medical Working Group was set up to draw up good practice guidelines and several million pounds were made available to develop drug services. Alternatively, its impact may have lain in re-involving the medical generalist,[59] albeit with strict controls, and in bringing drug services out of the hospital setting.[60] Spear saw the report's emphasis on a multi-disciplinary approach beyond prescribing as heralding the end of the dominance of hospital-based treatment services[61] but this was probably over-stating the case, given the subsequent difficulties in recruiting GPs to take up the challenge.

Most significant in the public-private debate was the raft of regulatory measures concerning prescribing. Spear noted the importance of the report's Chapter 7 which he saw as 'little more than an elaboration of the consultants' views' and an opportunity for 'the more politically motivated and forceful members' of the London Consultants Group 'to regain the influence they feared they were in danger of losing.[62,63] While the evidence supports Spear's argument, the consultants were not acting alone.

T&R considered it preferable for both NHS and private doctors working outside hospitals to liaise closely with hospital specialists and members of other disciplines in making their prescribing decisions. It also suggested that further knowledge could be gained by GPs taking up clinical assistantships in hospital-based services. Along with the other methods of surveillance and monitoring recommended by the *T&R* Working Group, these proposals could have enabled control of the prescribed drug supply to have been taken along the lines favoured by the London Clinic establishment represented by Bewley and Connell.

These two London consultant psychiatrist members of the *T&R* Working Group supported a very restrictive prescribing policy and opposed maintenance on opiates, especially outside the hospital setting. They favoured abstinence-oriented treatment over longer term prescribing, and methadone over heroin. The other two, Drs Parr and Thorley, had been invited onto the *T&R* Working Group at the end of 1978 by the medical civil servant Dr Sippert as 'permanent expert witnesses' due to their experience of treatment outside London (in Brighton and Newcastle respectively),[64] and, in the case of Dr Thorley, to counterbalance the London/South East dominance of the group.[65]

Connell had established his reputation with a study proving the previously unknown psychotic effects of amphetamine.[66] Bewley had been one of the first psychiatrists treating drug users in England during the 1960s.[67] Although he rose to high office, Bewley was something of an outsider to the English establishment. Born into an Irish Quaker family with many medical members he was educated at Trinity College, Dublin and underwent his psychiatric training in Ireland. Despite Bewley's strongly espoused opinions, he had a disarming tendency towards self-deprecation and claimed to have been the only person to have applied for a job at the Maudsley Hospital four times, succeeding on the fourth attempt.[68] Connell was reportedly more boastful,[69] with a dominating personality.[70]

Among psychiatrists around the London Clinics, there were a range of views on the wisdom of maintenance prescribing. However, those in the most powerful positions, including Connell and Bewley, seemed to have been successful in imposing their views on the majority of others at meetings of the London consultants held at the Home Office. They also took an interest in the regulation of the profession. Philip Connell was the Royal College of Psychiatrists' representative on the GMC from 1979, and Thomas Bewley replaced him in 1991.[71,72] In 1980, Bewley had been responsible for the first published attack on private prescribing, suggesting that control of psychoactive drugs should be confined to licensed practitioners,[73] views repeated in *T&R*. A few years later he reported Ann Dally, the best known private prescriber of the 1980s, to the GMC.[74,75] Both Connell and Bewley were based at hospitals with large numbers of drug-dependent patients. Both were members of the ACMD, of which Connell was to become chairman in 1982, and both held the post of specialist advisor to the Chief Medical Officer on drug dependence at various times. Dr Bewley became President of the Royal College of Psychiatrists in 1984.

Although the report had initially proposed extending licensing beyond heroin and cocaine to other drugs, it was Dr Bewley who had suggested that prescribing outside the Clinics merited a separate chapter.[76] In discussions Dr Bewley had gone further still, suggesting that the extended licence should cover all drugs controlled under the Misuse of Drugs Act under classes A, B and C, not just opioids. This was something the London Clinics' consultants had proposed to the government back in 1968.[77] Dr Thorley thought that a wider extension to non-opioids was too radical to receive practical support[78] and it was never recommended.

While these ideas *had* been brought to the forefront by Dr Bewley, interviews and committee documents show that there was consensus across the *T&R* Working Group that private prescribing needed to be tackled, even among non-medical members like David Turner.[79] Anthony Thorley recalled 'There was a real sort of keenness to try and tidy up the bad practice that existed in the private sector... So there wasn't a difference with Dr Bewley and the rest of the group I think on that one at all... And in fact, in a kind of way, I think it's quite reasonable to consider that one of the bedrock themes of the Treatment and Rehabilitation Group was to address this problem.'[80] Members from various professional backgrounds with different agendas were united in their agreement over the problem but there were different views on how it should be done. An early draft of Chapter 7 suggested that private doctors and GPs should only be granted licences to treat drug users if they worked with consultants in the Clinics. Although this idea was raised in the published chapter, it instead merely recommended close liaison with hospital services and access to expert second opinions. The call for reforms to the Home Office's Tribunals system was also introduced in the new chapter.

When considered by the ACMD, the new chapter on 'Prescribing Safeguards' elicited opposing views. Some council members felt that the report was too critical of private prescribers: although some private practitioners had misused their powers to prescribe, so too had NHS practitioners. 'Others pointed out that when patients were paying for prescriptions for drugs of addiction... there was more potential for abuse.'[81] However the ACMD did not wish to change the text. *T&R* itself avoided taking a stand on maintenance prescribing because, it said, expert opinions differed and decisions depended on individual circumstances. This would seem to limit the scope for producing consensus guidelines on good prescribing practice, yet the report urgently called for 'an authoritative statement of good practice, which should incorporate the need to make use of the support facilities we have mentioned'. Reference to 'support facilities' was another nod to the importance of the Clinics.

Although there was accord over the need to curb private prescribing, tension can be discerned within the group over whether non-medical members could make policy about the details of treatment. When David Turner had raised questions about disparities in the Clinics' prescribing practices, Dr Bewley commented that there was a problem of appearing to interfere with doctors' clinical freedom by making recommendations about treatment and whether to prescribe or not.[82] One medical member recalled agreement on the *T&R* Working Group, that 'the overall view around the table in Treatment and Rehabilitation was to see people come off drugs and that the idea of encouraging or in a sense, affirming their right to have long-term for life prescribing was not on,' but despite holding definite opinions the group 'was shy of itself making a strong statement about treatment... it wasn't really the business of the Working Party.'[83] Hiving off the production of the good practice guidelines to an all-Medical Working Group, as *T&R* recommended, prevented those outside the medical profession from 'interfering'.[84]

The establishment of the Medical Working Group also served another function. Those London psychiatrists who were against maintenance prescribing succeeded in moving discussion of the details of treatment content to an arena in which they were supreme. In the highly stratified world of medicine, Connell and Bewley, as the more experienced specialist hospital consultants, held seniority; if such an all-Medical Working Group were set up to deal with this matter separately, their views would carry the greatest weight. In the event the group was chaired by Connell with Bewley as a member and their anti-maintenance approach was victorious (see Chapter 3).

The guidelines recommendation seems to have been a compromise. Some of the *T&R* Working Group's psychiatrist members were pushing for statutory controls on prescribing to restrict drug treatment to the NHS and end private doctors' involvement. Opposing them was David Turner, who, like Bewley and Connell, was concerned about private doctors' prescribing but saw a danger in the Clinics holding a monopoly of treatment. Describing the Clinics' uniform approach to treatment, he recalled:

> The sense was that there was no clinical judgement involved once the decision was taken to prescribe. Everyone was to get their regulation dose of oral methadone without the inconvenience of having to take the individual situations into account. The client's task was to adapt to the 'treatment', the clinician had no responsibility to adapt treatment to the client.[85]

The secretariat, and in particular the DHSS's medical advisor on drugs, Dr Dorothy Black, also supported a wider range of treatment choice than was being offered by the Clinics, and may have helped to broker this compromise.[86] The guidelines could offer a deterrent to private over-prescribers without recourse to the law. Considering their subsequent use, it is ironic that the idea for guidelines was probably borrowed from the Association of Independent Doctors in Addiction (AIDA), a group of NHS and private doctors working outside the Clinics, led by Dr Ann Dally. AIDA had produced its own draft guidelines in 1982, on which Dorothy Black had provided comments, and these were circulated to the *T&R* Working Group that year.[87,88] *T&R*'s call for Home Office licensing to be extended to cover all opioids, as Bewley had recommended in 1980,[89] could also have effectively shut private doctors out of treating drug users, if licences had only been granted to doctors working in the Clinics. However, some members of the *T&R* Working Group did not object to the recommendation as they thought it unlikely to be implemented.[90]

T&R contained some interesting contradictions regarding 'good practice' in treatment. It recommended the preparation of an 'authoritative statement' on good medical practice[91] but had reservations about the feasibility of this. At one point the text reconsidered what it saw as the second Brain Committee's dilemma 'as to how far it was right to offer drugs to addicts as an inducement to seek or maintain treatment', and answered accordingly 'We do not consider...that there can be any simple answer to the question since expert opinions differ and much must depend upon individual circumstances. Rather we prefer an alternative, more flexible approach responsive to the varying problems faced by drug users.'[92] These apparently opposing views may have represented not just differences among the range of professionals but divisions among the medical members.

Aside from these conflicts of opinion, the report also conceded the limited research base on which treatment advice could be based: 'It is not possible...on the basis of research undertaken so far to demonstrate conclusively that one approach [to treatment and rehabilitation] is more effective than another.'[93] Then, rather surprisingly, the report declared that 'there has always been a broad consensus as to good and effective treatment of problem drug takers' but 'it has not always been widely known or widely applied'.[94] This varying range of views pointed to divisions within the *T&R* Working Group over the content of treatment over which 'there was clearly going to be no agreement'.[95] Minutes from an ACMD meeting that approved *T&R* also suggested a split on the Council over the prospects for producing good practice guidelines. Members spoke both of the 'diametrically opposed views on treatment' among experts, making agreement on guidelines difficult, but also 'a pattern of good treatment practice which it was hoped would emerge in discussions'.[96] Like the *T&R* Working Group, the ACMD itself seemed to have been divided over issues of maintenance- and abstinence-oriented treatments.[97]

Divisions can be seen within the *T&R* Working Group's psychiatrists along generational and regional lines when it came to the idea of incorporating other disciplines and professions into addressing the problems of drug use. The younger psychiatric consultant Dr Thorley, who had carried out research with sociologist Gerry Stimson, encouraged the report to take a multi-disciplinary approach and wanted to broaden understanding of the non-medical aspects of drug use. According to one member, Dr Thorley was 'of a newer generation, more open to working with other people and other services and keener on the idea of multi-disciplinary working...he represented a different approach and one not always welcomed by his consultant colleagues on the Working Group'.[98]

This chimed with Thorley's own view of himself, making a comparison with Thomas Bewley: 'The whole of the process was on the threshold of a, a rather different view looking at so-called drug addiction, which a number of us were quite keen in framing, sort of, new way of thinking. And he [Bewley] represented a kind of old school medical model, you know, in a very clear and identifiable way.'[99] Bewley was the most senior medical member of the committee and in the highly stratified system of medicine this could have an inhibiting effect on other doctors on the *T&R* Working Group.[100,101] Dr Thorley explained,

> [Dr Bewley] had a lot of personal influence, and power and so on. I mean he went on to take high office in the Royal College of Psychiatrists later, and so on and so forth, and he was very actively on the General Medical Council ... And, and so, you know, when you're just a young baby consultant coming along, and you've got somebody as senior as that in the medical kind of hierarchy, it's not easy to make a sort of, a, you know, start to initiate what was ... a bit of a paradigm shift really.[102]

The Conservatives' manifesto of 1979 pledged the government to simplifying and decentralising the NHS. Instead, as historian Charles Webster has noted, in the health service and elsewhere, the government actually ended up introducing a much greater degree of central supervision over local activities.[103] The manoeuvres behind *T&R* illustrated this process; with centralising recommendations emanating from the *T&R* Working Group, opposed by civil servants in the belief that they were 'upholding Ministers' policies' of decentralisation,[104] who were in turn overruled by ministers wanting to control matters from the centre.

T&R's recommendation for expanding central government's arrangements to advise and support local agencies was initially rejected by a DHSS official drafting the government response as it would 'conflict with Government policy on non-interference with local decisions on the allocation of local resources'.[105] Kenneth Clarke, then Minister for Health, was 'not very impressed' with this draft response and, in a memo to his boss, Norman Fowler, then Secretary of State at the DHSS, complained that 'Leaving the provision of service to "local decision-makers" will not make much progress unless we give them a steer.'[106] Furthermore, the government was also already committed to spending 6 million pounds centrally allocated on developing drug services on the *T&R* Working Group's recommendation.

T&R made very little reference to published research evidence. Only the three pages concerning 'The effectiveness of treatment and rehabilitation' mentioned a handful of studies. Statistics from the Home Office on the number of patients in treatment, drug offenders and drug seizures were used in the report, but most of the evidence used by the *T&R* Working Group was of a more informal type, derived from the experiences of its members and their visits around the country. These trips provided the opportunity for discussions with a wide range of workers in contact with drug users and with patients, ex-addicts and other concerned individuals. The *T&R* Working Group also took oral evidence at meetings; its concerns about doctors working outside hospital-based services was based on discussions with the Home Office Drugs Branch Inspectorate, doctors from the DHSS, doctors working with drug users and views expressed in medical journals and elsewhere.[107]

The lack of cited research evidence seemed in part to be a result of its limited availability at that time, the central point made in the report's chapter on research and confirmed elsewhere.[108] However, other reports by the ACMD published during the 1980s on topics for which there was much more research evidence, such as HIV/AIDS,[109] also lacked citations, relying again on submissions to the committee from organisations and individuals. The fact that the ACMD published reports during the 1980s without perceiving a need to support its statements through reference to published research implied a reliance on its authority as a body. 'Expertise' resided in its committee members' experience and assumed impartiality with an expectation that their conclusions could be trusted and that the information from which they were drawn did require independent scrutiny. This approach was not uncommon in medicine before the advent of the 'evidence-based medicine' movement but in such a politicised field the assumption of an objective, neutral expertise, whether truly possible in any circumstances, was particularly open to abuse.

T&R's implementation benefited from Norman Fowler's interest in drugs, which dated back to his time as a journalist before entering politics, and public awareness was growing with the dramatic increase in Britain's drug use.[110,111] In turn the government was keen to gain the support of the medical profession for the report's proposals and called a conference of medical representatives in January 1983 to achieve this. In a departure from the normal protocol, Norman Fowler gave the keynote address, a job normally left to a more junior minister or senior official,[112] again reflecting the new priority given to drugs policy.

Overall the minister seems to have received a positive response from the medical representatives who showed no greater sympathy for private prescribers than had the *T&R* Working Group, with reports from the conference of some disquiet about 'the principle of private prescribing' in view of the charging of fees.[113] There appeared to have been dissent as to whether GPs should treat drug misuse but agreement that if they were to be, then training and additional support would be required.[114] The British Medical Association's General Medical Services Committee, representing the majority of GPs, supported the recommendations for good practice guidelines and for extending licensing initially to dipipanone and later other opioids,[115] as did the majority of medical representatives.

The requirements for doctors to obtain licences proposed at the January 1983 medical conference were strict: those who wished to be able to prescribe methadone and other opioids by Home Office licence should have additional training, multi-disciplinary support and have membership of the Royal College of General Practitioners or the British Medical Association.[116] These obstacles to practice may have reflected the general lack of enthusiasm among GPs for treating drug users. Another suggestion was that a GP should be limited to treating only three or four drug-dependent patients, effectively ending private prescribing on any significant scale. The good practice guidelines proposal was more controversial and the DHSS agreed to invite further small groups to consider both the question of licensing and the preparation of guidelines in the light of responses to a wider consultation exercise.[117] A civil servant's draft of Norman Fowler's letter to the Home Secretary gave a rather more triumphant tone to Fowler's achievements at this meeting where he 'secured a favourable climate' for the establishment of the good practice guidelines and licensing working group.[118] In the actual letter Fowler sent, mention of the 'favourable climate' had been removed.[119]

Conclusion

Treatment and Rehabilitation heralded many changes to the freedoms and responsibilities of doctors working outside the Clinics and the first major regulatory interventions against private doctors since the second Brain Committee. While attacking private doctors, *T&R* also gave approval to re-involving generalists in the treatment of drug users, reversing over a decade's policy of exclusion. Such expansion and concerns over existing non-Clinic prescribing were used to justify the retention of power for the hospital consultants and central government through the development of new and existing control mechanisms. These controls were in fact primarily designed for existing private prescribers rather than any anticipated GP involvement. That the report's recommendations were implemented almost wholesale can be attributed to widely publicised changes in the landscape of drug taking in Britain since the late 1970s and the political will to visibly address these. Some of these wide reaching changes might never have been suggested or given such prominence had it not been for the determination of a few individuals deeply concerned about the role of private prescribers and perceived encroachments on their dominant position.

Although expanding, the small size of the drugs policy community in the late 1970s and early 1980s allowed certain ambitious actors to gain great influence across a number of settings. Philip Connell and Thomas Bewley's authority within the London psychiatric drugs field and their presence on this and subsequent working groups played a pivotal role in attempts to control private prescribing. Although opposed to the monopoly over controlled drug prescribing that Clinic consultants sought, concerns about private prescribing struck a chord with the voluntary sector representation as well. Yet these powerful actors were not able to get their way entirely and needed the support of civil servants and politicians to succeed on each policy.

While the second Brain Committee had been entirely medical, in the day of the *T&R* Working Group medicine was having to make room for other disciplines and occupational groups. The growing voluntary sector, representing a more social and less medical model of drug use, gained greater recognition – as seen when David Turner set the second *T&R* Working Group's radical agenda. By 1982 drugs had begun to assume a higher political profile, attracting renewed ministerial interest. Yet doctors successfully defended their territory from infringements and managed to keep the most controversial treatment issue – prescribing – within their professional borders. The story of what happened within those borders is told in the next chapter.

3
Defining 'Good Clinical Practice'

There was no question of a really serious long-term option of prescribing forever... That's something we were actually trying to stop... Because all over London there were these geriatric junkies to put it very rudely, people who had been prescribed out of the Sixties... a rump of people who have just never changed... So, rather than again give these people in London, the London Harley Street stuff, [the private prescribers] a kind of green light to go on prescribing forever, we decided to have it self-limiting... And, of course there were three-month and six-month so-called detoxifications, which we did use in Newcastle. And, I mean they were reasonably successful but of course this was all anecdotal.[1]

Introduction

Hostilities between private prescribers and their critics did not cease with *Treatment and Rehabilitation*. Attacks from each side continued in medical journals[2] and carried over into a new arena of medical regulation: guidelines on good clinical practice. Published in 1984 by the Department of Health and Social Security (DHSS),[3] these first official guidelines in British medicine were followed thick and fast by many more, particularly after the NHS market reforms of 1991. Some have represented this as an increase in state control over the medical profession and a weakening of doctors' autonomy.[4] However, this chapter argues that these guidelines embodied the use of regulation by an alliance of one part of the medical profession with an arm of the state to control the practice of a second group of doctors.

The *Guidelines* were used to secure the ascendancy of one particular treatment model and impose this on all doctors, while citing no supporting published research evidence. Once again, the experience of an expert committee was deemed sufficient by government and many of those involved for determining 'good practice'. As discussed in the previous chapter, two years earlier the Advisory Council on the Misuse of Drugs (ACMD) had published *Treatment and Rehabilitation* and its recommendations had started to change the direction of drug treatment policy in England.

At that point, doctors working outside the Clinics were still able to prescribe methadone, a synthetic opiate used to replace heroin, dexamphetamine (a stimulant of the amphetamine family) and other substitute drugs. Their prescribing was receiving unwelcome attention, particularly from senior consultant psychiatrists in the London Clinics. Chief among those irritated by the private prescribers were Dr Thomas Bewley and Dr Philip Connell. They had encouraged the move from maintenance heroin prescribing to short-term methadone detoxification and from injectable to oral formulations across the London Clinics.

In the 1970s and early 1980s little research had been carried out to evaluate these different approaches to prescribing,[5] and what existed was often misrepresented. Richard Hartnoll and Martin Mitcheson's randomised trial of injectable heroin and oral methadone was frequently cited to justify the prescribing changes in the Clinics and had been key in supporting the move from maintenance prescribing of heroin to limited stabilisation on methadone followed by short-term detoxification, often with obligatory therapy sessions.[6] In the early 1970s Richard Hartnoll and Martin Mitcheson randomly allocated 96 opiate-dependent patients at a North London Clinic to either injectable heroin or oral methadone maintenance treatment and then followed up a year later.[7] The research was carried out between 1972 and 1976 but was unpublished until 1980.

In their published paper the study's authors were equivocal about its findings, stressing they showed no one treatment to be superior. Although their results showed different positive and negative points for both the heroin and the methadone prescription groups, 'the differences between the two groups, although often statistically significant are not startling. Whichever treatment is given, there are obvious casualties that may reflect the pre-existing chaos of the patients as much as the treatment offered'.[8] They concluded that the findings 'contribute to a more informed discussion' of the issues around heroin prescription 'rather than provide an unequivocal answer'.[9] Yet in spite of these cautious words, the research had already been used to support the switch from heroin prescribing and towards oral methadone across the Clinics.[10] Martin Mitcheson, co-author of the study and consultant at University College Hospital's Clinic, stopped prescribing injectable drugs entirely to his new patients after the research was completed in the mid-1970s.[11]

Thomas Bewley described how, because the evidence showed neither drug to be superior, 'I felt it was open to the prescriber to choose so I moved over to methadone and phased out heroin',[12] trying to encourage other doctors to follow suit. One of the opponents of the Clinics' switch to methadone complained that 'while critics of what the [Clinics] were doing were required to produce data to support their criticisms', one of its proponents had freely admitted that there was 'no scientific basis' for this major change.[13] By the time the *Guidelines* were being drafted in 1984, oral methadone detoxification was the only option offered to new patients seeking help from the Clinics.

According to John Strang, a senior London Clinic psychiatrist and prolific and influential researcher who became one of the key players in the control of prescribing from the late 1980s, there developed a 'therapeutic apartheid' between these new patients and those who had attended the Clinics pre-1975 who often still received maintenance supplies of injectable drugs.[14] The near monopoly of treatment the Clinics held, their leaders' willingness to conform to a single model, and the absence of strong patient voices had allowed the Clinics to become unresponsive to the preferences of their patients; while the private doctors, practising on a more consumerist model, were able to supply unmet demand. Although the voluntary sector had been providing drug services for many years, these were typically not medical and did not prescribe.

The London Clinics' unified approach was corralled and reinforced by the regular meetings of their consultant psychiatrists, from 1968,[15] in order to share information and standardise practice. These were held initially at the DHSS and from 1977 at the Home Office.[16] Martin Mitcheson described these meetings as 'typically English, discreet peer group pressure tending to moderate the prescribing of heroin' in order to prevent drugs being traded illegally.[17] One psychiatrist who continued to disagree with the anti-maintenance approach at the London consultants groups claimed to have been pressurised to conform when he persisted in the practice.[18] Another complained that these conformist pressures produced farcical double standards: psychiatrists who continued to prescribe injectable heroin were criticised but licensed colleagues in other Clinics would phone to ask them to prescribe heroin to a patient because, although licensed, they did not feel able to do so themselves.[19] As mentioned earlier, doctors wanting to prescribe heroin or cocaine for the treatment of addiction (and from 1984 dipipanone as well) had had to apply to the Home Office for a special licence. The licences were almost exclusively granted to psychiatrists working in the new NHS Clinics; only two or three doctors were ever licensed to prescribe heroin privately.[20,21]

As discussed in the previous chapter, drug users were increasingly seeking treatment from NHS or private general practice at the end of the 1970s. This was happening amid a shortage of treatment places in the Clinics and heroin use spreading to parts of the country without specialist provision.[22] The move away from the Clinics may have resulted not just from their long waiting lists but also because of their changing prescribing policies,[23] particularly in London. Clinic psychiatrists expressed disquiet at these changes, particularly over private doctors prescribing outside the Clinic system on a fee-paying basis.

The peer pressure exercised successfully by the London psychiatrists continued after heroin prescribing had been curtailed but it failed to impose conformity on the practice of private prescribers or GPs. These doctors prescribing outside the Clinics greatly valued their independence from their peers. As independent contractors to the NHS, and with so few of their number apparently interested in treating drug users, there was no equivalent attempt among the ranks of GPs to establish a clearly defined approach. Among the more patient-led private doctors who catered to needs or desires unmet by the NHS, there was greater sympathy for more liberal prescribing and less concern about pressure from within the medical profession. Private prescribers did not require high status or position to continue to maintain a good income from the treatment of drug users. The main theoretical threat to their livelihood was from disciplinary action by the Home Office Drugs Inspectorate or the General Medical Council which could stop such prescribing, but this was relatively rare at this point. While the members of the Association of Independent Doctors in Addiction (AIDA) had tried to agree some criteria for good treatment by producing their own guidelines, ultimately they had 'agreed to differ' and their guidelines were never finalised (see Chapter 6).[24,25]

Conservative politicians who might be seen as champions of private treatment for drug users did not involve themselves in this particular debate. Although the new Conservative government of 1979 had greatly facilitated the supply of consultant labour to the private sector,[26] private prescribing for the treatment of addiction by NHS Clinic psychiatrists was rare and not well respected. The 1979 Conservative manifesto had proposed an end to the 'vendetta' against private practice and its 1983 successor encouraged a positive role for private medicine.[27] However, this did not include drug treatment which was dealt with as a 'drugs' issue rather than a 'private medicine' issue, and as such was not considered a party political concern.

Home Office Drugs Inspectorate

Unlike other areas of prescribing which the medical profession regulated itself through the General Medical Council, the prescription of controlled drugs also came under the scrutiny of the Home Office. These powers dated back to the First World War[28] and developed through the Dangerous Drugs Act 1920, into two different systems of monitoring: the Home Office's own Drugs Inspectorate concerned with 'irresponsible prescribing' and the police's chemist inspecting officers concerned with criminal offences. 'Irresponsible prescribing' was never defined by law and up until publication of the *Guidelines*, the Home Office had no official measure against which to gauge it.

Much of the Home Office Inspectorate's regulatory work was carried out on an official but informal basis, with inspectors visiting doctors and advising them to modify their practice.[29] On rare occasions, doctors considered to be prescribing irresponsibly were summoned to a Home Office Tribunal, a panel of doctors that could recommend to the Home Secretary that his or her controlled drug prescribing rights be removed if found guilty of irresponsible prescribing.

H. B. 'Bing' Spear, who had been active in the Inspectorate since 1952, possessed an intimate knowledge of the drugs 'scene', taking a personal interest in the prescribing habits of doctors and in the well-being of individual drug users. He became not merely an implementer of others' policies but a major influence in his own right. Moving as he did among the doctors, civil servants, committees and drug users on the streets, he was recognised by all sides of the prescribing debate as one of the most knowledgeable and trustworthy sources of information and guidance.[30–32]

In the tradition of British civil servants, he was careful not to appear partisan and concealed deeply held views which were only expressed publicly after his retirement.[33] Although a strong supporter of doctors' freedom to prescribe on a maintenance basis and an opponent of the changes brought in by the London Clinic psychiatrists, he also believed in the need to regulate doctors' prescribing.[34]

Until ill health forced him to retire in 1986, Spear attended most of the London Clinic psychiatrists' meetings,[35] where he was able to provide information to the consultants and in his regulatory capacity and could follow up reports of irresponsible prescribing among other doctors.[36] Tribunals for irresponsible prescribing were never brought against doctors working in the Clinics,[37] who were free to prescribe within the standards they set for themselves. For them, the picture was one of self-regulation rather than regulation by the state.

Membership and intentions

At the *Treatment and Rehabilitation* Working Group, the medical profession had successfully preserved prescribing policy for themselves. Encouraged and facilitated by the DHSS and its medical civil servants, the *Guidelines'* membership reflected this. Most of the all-Medical Working Group's members had been nominated by medical bodies at the invitation of the DHSS: the General Medical Council, British Medical Association, the Royal Colleges of Psychiatrists and General Practitioners, the Joint Consultants' Committee and AIDA.

This last group was led by the outspoken Ann Dally. With the encouragement of Bing Spear who was sympathetic towards private practice and maintenance prescribing, Dally and other private and NHS drug doctors outside the Clinics had set up the Association in 1981. They attempted, ahead of the field, to produce their own good practice guidelines and other policies to raise standards and self-regulate. In a deliberate political move the *Guidelines* chairman and secretariat included AIDA to create at least the appearance of a consensus statement.[38,39] Views vary as to whether there was a genuine intention to take on the views of private doctors or in fact an attempt 'to smother the enemy . . . by creating something they appear to agree with'.[40]

Professor Neil Kessel, in his role as the Chief Medical Officer's advisor on alcohol, was appointed to the Medical Working Group and a minority were invited for their particular expertise: Dr Arthur Banks had written on treating drug users in general practice,[41] and Elizabeth Tylden was an authority on drug use in pregnancy,[42] creating a mix of representation from medical bodies and expertise in the drugs field from across NHS and private medicine.

The original aims of the psychiatrist members of the *Treatment and Rehabilitation* Working Group and the chairman of the Medical Working Group, Dr Philip Connell, in producing the good practice *Guidelines* were to: control doctors working outside of the Clinics, particularly those in private practice;[43,44] retain dominance for drug dependence psychiatrists and their preferred treatment model; and prevent diversion of prescribed drugs onto the black market. The first papers circulated to Medical Working Group members were an article criticising private prescribing and related correspondence in the *British Medical Journal* and *The Lancet*.[45,46]

Thomas Bewley had similar motives to Dr Connell: stopping maintenance prescribing and promoting the model of treatment dominant among London psychiatrists. Dr Bewley, at this time, had been won over to methadone from heroin prescribing after meeting Vincent Dole, a pioneer of methadone substitution therapy, during a visit to the USA in 1967.[47,48] He also hoped to put a stop to private prescribing.[49,50]

Arthur Banks, a GP in Chelmsford, Essex, was experienced in treating drug users and wished to encourage other GPs to get involved. There was little published guidance available to GPs at that time and he hoped that the *Guidelines* would give them greater confidence and show their obligations in treating drug users. He wanted 'something official that was a considered document summarising the best ideas on treatment of drug addicts and something that would be available to all GPs'.[51] He did not have specific concerns about the potential for the *Guidelines* to enforce a particular model but strongly opposed the extension of licensing to all opioids, seeing it as likely to destroy any emergent interest in treating drug users from general practice.[52] He also wanted to show GPs that there was government backing for their involvement independent from addiction psychiatrists.

Ann Dally also opposed the extension of licensing and wished to promote her views on treatment, including the need for long-term prescribing. Similar to virtually all doctors practising outside the Clinics, she did not have a licence to prescribe heroin or cocaine. A fellow member recalled that she 'fought her corner with great vigour'.[53] Had the licensing system been extended to cover all opioid drugs, she might have been denied the right to continue prescribing to most of her patients. She saw herself as one of a group of 'dissidents' which included Dr H. Dale Beckett whom she had invited onto the Medical Working Group from AIDA, and sometimes Arthur Banks, opposing the psychiatric 'establishment' on the committee.[54,55] Psychiatrist Dale Beckett, at this stage retired from his NHS consultant post in charge of a Clinic at Cane Hill Hospital, Surrey, and working in private practice, held unorthodox views on the rights of drug users to maintenance supplies, believing that heroin, a 'gentle drug', should be made available to addicts and he supported Ann Dally on treatment and licensing issues.[56,57]

Among the representatives of the medical bodies without specific drugs expertise, the British Medical Association's J. A. Riddell, a Glasgow GP, strongly opposed GP involvement in treating drug users.[58] 'He just felt they'd be overwhelmed; there'd be more problems because they wouldn't cope',[59] and on most issues the other non-expert representatives tended to side with the psychiatrists, including on the matter of licensing.[60,61]

Within the Medical Working Group, the push for more restrictions on prescribing through Home Office licensing, which would have considerably reduced some doctors' clinical autonomy, came from the psychiatrist members appointed by the secretariat. It was supported by many of the elected doctors and opposed by medical civil servants at the DHSS and administrative civil servants at the Home Office.[62,63]

Why then, would some doctors try to restrict their own profession's autonomy? Although it was unknown how exactly such a licensing system would have operated, most likely it would have fitted into the existing scheme for heroin and cocaine prescribing. The Clinic psychiatrists already held these Home Office licences and were not afraid that their powers would be affected.

The secretariat

Dr Dorothy Black, senior medical officer responsible for drugs policy at the DHSS, and Mr R. Wittenberg, a career civil servant, were secretariat to the Medical Working Group. Dr Black had come to the DHSS in 1981 from her post as consultant psychiatrist working with drug users in Sheffield. During her time at the DHSS she was particularly influential in drug treatment service policy. Despite sharing a medical specialty, she did not automatically side with the dominant London addiction psychiatrists and encouraged non-statutory and non-medical involvement in treatment services.[64] Her experience of patterns of drug use outside of London was important in countering the London-centric policy-making of the period.[65]

One member of the *Treatment and Rehabilitation* Working Group remembered Dr Black as proposing guidelines as a compromise between the London consultants' call for legal regulatory changes and those opposing them. 'Dorothy Black was important... in avoiding formal regulation in favour of guidelines... Partly I think that regulation was a rather impractical process but secondly I think that Dorothy was more conscious of the need for a greater range of treatment options rather than a very standardised system [of the Clinics].'[66]

Her wide scope to initiate policy in the DHSS of the early 1980s was facilitated by both the lack of interest in drug treatment policies among the administrative civil servants (i.e. those who were employed as career bureaucrats rather than hired for their particular expertise in a subject) and the enthusiastic support of Norman Fowler, then Secretary of State.[67] According to a contemporary source in the DHSS, 'there was nobody else in the Department who knew anything at all about drugs... from the point of view of the administrative civil servants it was almost seen as being sent to Mongolia'.[68] Dr Black was closely involved in the selection process for membership of the Medical Working Group and carried out most of the *Guidelines* drafting work as Mr Wittenberg was unwell for much of the project.[69]

An important feature of the DHSS was the inclusion of medical civil servants on its staff directly answerable, until 1995,[70] to the Chief Medical Officer. These medical civil servants acted as 'experts' and tended not to become fully assimilated into the bureaucracy. Compared to the administrative staff they had considerable independence[71] to work as 'professionals' and played a significant part in initiating policy, which was then carried out by the administrative staff.[72] For instance, as DHSS 'observers' on ACMD Working Groups, they were encouraged to speak as experienced clinicians rather than administrators,[73] and many working in drug and alcohol policy returned to clinical work after periods at the Department during the 1980s and 1990s. In addition to the secretariat who drafted the document, observers from both the DHSS and Home Office attended the Medical Working Group's meetings, reflecting the Home Office's regulatory interest.

Regulation by the state

The *Guidelines* intended to help 'identify those cases where prescribing practices might be regarded as irresponsible'.[74,75] They were therefore valuable to the Home Office Drugs Inspectorate in their role of advising doctors and bringing Tribunal proceedings against them, helping the state use bureaucratic rules to control otherwise self-regulating professionals. As one DHSS civil servant commented, 'The Inspectorate, if you like, had their own internal view [on prescribing] and there'd never been any guidelines before... it would reinforce what the Inspectorate said as they trundled round all the doctors.'[76]

Not only were the *Guidelines* official, but they gave the appearance of medical self-regulation, rather than regulation by the state. For the Inspectorate it would 'give them another piece of support when they were advising doctors, that... to a doctor it might be more effective in influencing their practice to say "This is from a working party of doctors", rather than saying as a Home Office inspector, "Do you think your prescribing levels are too▪high?"[77] Doctors' sensitivities over lay people commenting on their prescribing proved to be a recurring theme.

Perhaps surprisingly, the Home Office did not seek the extension of licensing which would have given it greater powers over prescribers. Bing Spear was suspicious of the leading psychiatrists' intensions in their attempt to extend licensing and later accused Philip Connell of paying 'lip service... to the concept of clinical freedom' while 'conformity and psychiatric domination of the drug misuse field remained the ultimate goals'.[78] Here was a department of the Home Office acting within the policy community in part to its own agenda. Spear was concerned to control the flow of prescribed drugs from reaching the black market which was part of the Home Office's remit for regulating doctors but he also strongly believed in the traditions of the 'British System' and the freedom it allowed to prescribers. Under his leadership, the Inspectorate allied with the London psychiatrists' interests of producing the *Guidelines* when seeking to reinforce its own policing powers and opposed them on policies he saw as too restrictive. With no part of the drugs policy community strong enough to push through the policies they wanted alone, each faction made opportunistic and temporary alliances to achieve them.

What the *Guidelines* said

In style and presentation the *Guidelines* were functional and unembellished. Short paragraphs of impersonal, detached text gave an impression of authority and consensus and a sense that treating drug users was straightforward and relatively simple with limited variation. The content addressed the various doctors both inside and outside hospitals who might be involved in treating drug users, including GPs, psychiatrists and casualty officers. It focused on opiate, barbiturate and benzodiazepine dependence with just a few sentences on alcohol, stimulants and other drugs.

The *Guidelines* told all doctors, including GPs, that it was their duty to provide care for their patients' drug-related problems. Abstinence and cessation of injecting were the goals of treatment; long-term opiate prescription was strongly discouraged and GPs were told to consider it only under the guidance of a specialist. The substitute drug of choice was oral methadone to be used only for withdrawal over no more than six months. Patient and doctor needed to agree the detoxification regime (but this was in an absence of any alternative prescribing). No concessions to injectable prescribing were made, although patients' dependence on 'injecting and injecting practice' was acknowledged. Doctors were also advised to consider prescribing non-controlled drugs instead of opioids to alleviate withdrawal symptoms.

An appendix 'Managing withdrawal symptoms and detoxification' set out various detoxification regimes for use inside and outside hospitals. Here too most attention was devoted to opioid, barbiturates and benzodiazepine dependence. For opioid withdrawal, no limit was set on the dose of methadone that could be prescribed but the suggestion was that doctors were unlikely to need to prescribe more than 80 mg a day. Prescriptions of 80–100 mg of methadone, the *Guidelines* advised, should not be attempted in outpatients (i.e. by GPs or private prescribers). The 1999 *Guidelines* recommended a daily limit of 120 mg.[79] Prescribing regimes ranged between two weeks, which required the patient to be in stable accommodation and to receive intensive support from the doctor, and family or friends, to up to six months for which domestic stability was also needed. Daily dispensing to ensure the methadone was consumed only by the patient was encouraged.

As with other reports on drug treatment from the mid-1970s onwards, the 'multi-disciplinary' approach was advocated, both in hospitals through team working and by liaison with other agencies.[80] The *Guidelines* advised GPs not to manage more chaotic patients or those on high doses but rather to refer them to hospital-based services. In short, doctors were advised on the range of prescribing they should and should not undertake, the type of patients they should take on or refer, the context in which they should prescribe, the acceptable drugs, doses, and formulations and their duties to drug using patients. This did not reflect agreement across the views of the Medical Working Group or the summation of research findings but rather the dominance of certain doctors' views over others.

Licensing

The introduction of special Home Office licences for prescribing heroin or cocaine for the treatment of addiction in 1968 was quickly followed by a series of attempts, originating with the London Clinic consultants, to extend these requirements and restrict further the prescribing powers of doctors outside the Clinic system. The case was made for this through concerns about the diversion of prescribed drugs onto the black market, blamed on doctors working outside the Clinics, and sometimes a disapproval of maintenance prescribing itself.

The initial attempt occurred almost as soon as the Clinics had been set up and the first heroin and cocaine licences issued. A Department of Health meeting of the London Clinic psychiatrists in 1969 had proposed that *all* dependency producing drugs to known addicts, not just heroin and cocaine, should be removed from GPs and limited to the Clinics. However, the idea was rejected by the Department for financial reasons.[81] It was revived at meetings between the voluntary sector and consultants in 1979 and 1980 aimed at providing recommendations for the ACMD's *Treatment and Rehabilitation* Working Group. The result was a recommendation that all maintenance prescribing should be reserved for specially licensed doctors who must work either inside, or in close co-operation with, a specialist facility. Although this recommendation seems to have misrepresented the views of a number of the voluntary sector agencies, it was forwarded to the ACMD nonetheless.[82] A slightly modified version appeared in *Treatment and Rehabilitation* calling for production of the *Guidelines* and the extension of licensing to cover all opioid drugs.[83] The Medical Working Group recommended by a vote of eleven in favour, one abstention and three opposed in favour of the licensing extension,[84] but it was rejected by DHSS and Home Office ministers as unnecessary and possibly likely to deter GPs from treating drug users.[85,86]

John Patten, Parliamentary Under-Secretary of State for Health, was reluctant to extend licensing because of the resource implications, fearing also that it could undermine government policy of encouraging GPs to accept more responsibility for treating drug users.[87] Before advising the Secretary of State he sought information on trends in opioid prescribing. It was an analysis of prescriptions for two drugs of particular concern, dextromoramide (*Palfium*) and dihydrocodeine tartrate (*DF118s*), which seems to have convinced Patten's successor, Ray Whitney, that these extra controls were not needed.[88]

However, in his letter advising David Mellor, his opposite number at the Home Office, against extending licensing, Mr Witney gave a partial and more optimistic interpretation of the prescribing trends than the actual figures allowed. Palfium prescribing had declined since 1978 as Mr Witney claimed but *DF118* prescribing had actually increased since that year.[89] Furthermore, any falls in the number of prescriptions for these opioids might have represented the privatisation of its prescribing, the very issue which had prompted calls for the extension of licensing in the first place.

The prescription analysis document had taken a sample of 1 in 200 prescriptions. No source was given for the data which could have come from either the controlled drugs registers held by pharmacists (including every prescription issued whether private or NHS but which were not held centrally) or from the predecessor body to the Prescription Pricing Authority which collected data on every prescription issued under the NHS. Given the sampling it is most likely the DHSS was using the centrally held NHS information therefore leaving private scripts out of the calculations altogether. After a series of correspondence in which other officials repeated Witney's misinterpretations of the data,[90,91] David Mellor announced that he was 'entirely content' not to extend licensing for the time being.[92]

There was no indication that the prescribing policies pursued by the psychiatrists were influenced by politicians' views. From the major changes of the late 1960s until 1981, prescribing policies were of very little political interest at the DHSS,[93] and the Home Office Inspectorate, while keeping a close eye on prescribing outside the Clinics, had made little use of the Tribunal system to discipline doctors. Ministers at the DHSS were not interested in the *Guidelines'* content but applied great pressure for their completion putting it above any consideration of licensing extension.[94] Evidence suggested that the politicians' motive was to expand treatment provision amid heightened public concern about heroin, rather than to control prescribing, as the psychiatrists had intended. The Conservative government presented the *Guidelines* as a plank in their response to the heroin epidemic of the early 1980s[95] to encourage greater involvement of the medical profession in the care of drug users and so increase treatment provision.

Winners and losers

Fulfilling the wishes of the London psychiatric establishment, the final version of the *Guidelines* declared that long-term maintenance prescribing was only to be undertaken by psychiatrists and GPs with specialist supervision experienced in that approach.[96] Although Thomas Bewley's first draft of the appendix was not used and Anthony Thorley, the young consultant psychiatrist working in Newcastle,[97,98] was responsible for the final version, Bewley's overall prescribing preferences were reflected and abstinence was to be the clear strategy.

The *Guidelines* acknowledged that 'few psychiatrists have any specific training or wide experience in the treatment of drug misuse' but considered it their responsibility to provide advice and support for GPs in areas where there was no specialist drug treatment unit. Another Bewley and Connell preference was reflected in the emphatically stated superiority of methadone over heroin. Even in hospitals 'there are no clinical grounds for heroin or any other opioid being prescribed' (except allergic reaction to methadone).

Arthur Banks was gratified by the *Guidelines'* initial statement that 'All doctors have a responsibility to provide care for both the general health needs of drug misusers and their drug related problems', going against the wishes of Riddell, who opposed GP involvement. Although those opposing the extension of licensing to all opioid drugs lost the battle in the Medical Working Group, they won the war when ministers rejected the proposal, probably on account of the Home Office's advice. In the *Guidelines*, the Home Office Drugs Inspectorate and the General Medical Council gained a new medically authorised standard for prescribing which could be used in their regulatory work.

On the losing side, AIDA expressed its views about the *Guidelines* the following year,[99] criticising the advice not to prescribe substitute controlled drugs before assessment was completed. Underlying AIDA's criticisms was an emphasis on the individuality of patients, the high likelihood of failure in detoxification, of drug dependence as a long-term problem, the suffering resulting from withdrawal symptoms and the need to take into account addicts' immediate need or desire for a prescription.

Perhaps reflecting the different power relationship between the private doctor and his or her patient, the patient was seen as determining treatment to a greater extent than in the NHS. For instance, AIDA criticised the Clinics saying that 'addict in-patients who are not given the drugs they feel they need and with whom no rapport is made will either have drugs smuggled in or will discharge themselves, regardless of their physical health'.[100] This encouraged more generous prescribing with the balance tipping towards the individual patient than to public health or concerns about controlling the supply of diverted drugs. In her autobiography, Dr Dally entitled this chapter of her career 'The Misguidelines'.[101]

Dally's autobiography alleged that she and other 'dissidents' were tactically outmanoeuvred when they wanted to issue a minority report expressing their opposition to the *Guidelines*. New committee procedures, she claimed, were introduced so that they were not allowed to register their protest.[102] This was confirmed by one of the consultant psychiatrist members who recollected a change in the committee rules sidestepping the need for final agreement. He attributed this to behind-the-scenes activity by the Chairman and secretariat.[103] While the secretariat was influential in terms of members' selection and committee procedures, its limitations were perhaps revealed by the content of the *Guidelines*, which were less liberal than might have been expected from Dorothy Black's approach to prescribing.

The ACMD's *Treatment and Rehabilitation* expressed the intention, probably originating with Bewley and Connell, that conforming to the *Guidelines* would be used as conditions for licences once the licensing system had been extended to cover all opioid drugs.[104] As this extension never took place, the *Guidelines* had less of a disciplinary role than originally intended. The London consultants did not admit defeat, however, and continued to re-introduce the idea of the licensing extension to government through the 1980s and 1990s. These events are explored in Chapter 7.

What would appear to be a simple provision of guidance from 'experts' to other professionals raised many questions about both motives and methods. Like the *Treatment and Rehabilitation* report, it revealed the problematic nature of 'expertise' and evidence in a polarised and highly politicised field, the ways in which the medical profession has regulated itself, and the roles of government.

Why impose one treatment model?

A question central to these political activities and considered by Stimson and Oppenheimer[105] and Spear[106] was why this particular group of psychiatrists placed so much importance on the universal adoption of their treatment model to the exclusion of all others. The lack of research to support the change in the Clinics' prescribing policies was conceded by some of the key psychiatrists involved,[107] and another psychiatrist member of the *Guidelines'* working party, and key advocate of its recommendations described how they had come into being.

> There was no question of a really serious long-term option of prescribing forever... That's something we were actually trying to stop... Because all over London there were these geriatric junkies to put it very rudely, people who had been prescribed out of the Sixties, were now into the Seventies, late Seventies, early Eighties, here we were, there's a rump of people who have just never changed, because, in a way they've never had sufficient sort of multidisciplinary support around them, and the sort of framework of prescribing to really encourage them to come off with the treatment and rehabilitation package that we'd been trying to advocate the year before... So, rather than again give these people in London, the London Harley Street stuff, [the private prescribers] a kind of green light to go

on prescribing forever, we decided to have it self-limiting...And, of course there were three-month and six-month so-called detoxifications, which we did use in Newcastle. And, I mean they were reasonably successful. But of course this was all anecdotal. Nothing was tested with double-blind clinical trial. Everything was really opinion. Which of course was dangerous at one level.[108]

The 1984 *Guidelines* themselves were not 'evidence based', nor did they claim to be. They have since been retrospectively legitimised by the evidence-based medicine movement, with heavily referenced editions in 1999, and 2007[109,110] but in their first (and second) incarnation were a summation of personal experience, hospital testing of treatment (not necessarily published) and various textbooks which might be the work of a single psychiatrist writing about his or her experience or what had been gathered from colleagues.

At that time, the personal opinions and experience of senior doctors was considered a suitable basis for 'good practice' and in this sense they did not appeal to an external body of 'scientific' data to justify their statements, as Harrison and Ahmad have described in later guidelines.[111] There was, in fact, little published research evidence on the efficacy of treatment at that time but what there was, such as the Hartnoll and Mitcheson trial comparing heroin and methadone prescription, was not mentioned in the *Guidelines*.[112] The first *Guidelines* contained no references to scientific studies, only reports, textbooks or reference sources such as the British National Formulary.

Part of the change in Clinic prescribing policies which so influenced the *Guidelines* can be attributed to the 'silting up' of treatment spaces with long-term maintenance patients and professionals frustrated at their lack of impact on their patients.[113] Yet extraordinary measures were taken including attempting to get doctors disciplined if they opposed the newly favoured abstinence-based approach. Struggles for prestige and status within the medical profession may explain this.

The new model of treatment (short-term methadone detoxification and no injectable prescribing) described in the *Guidelines* allowed psychiatrists to achieve change in their patients even if that change was short lived. Maintenance prescribing of injectable heroin, the drug that would have been used by the patient outside of treatment, could be seen as a passive professional approach, where any change in behaviour was initiated by the patient rather than the doctor. Clinic psychiatrists' preference for more 'active intervention',[114] where patients were given restricted options and required to sign contractual agreements, could be seen as an attempt to gain greater job satisfaction and prestige for their emerging specialty.

Stephen Shortell, writing in 1974, showed that the relative prestige of a specialty within the medical profession corresponded to the activity or passivity of the doctor in the therapeutic relationship. The more active the doctor was in relation to the patient, the higher the prestige of the specialty, with surgery, for instance, where the doctor would perform actively on the passive patient, scoring highly. The more the doctor relied on patient participation, and acted to 'help the patients help themselves', the lower the prestige.[115] Prescribing oral methadone instead of injectable heroin was seen as more 'confrontational',[116] therefore more therapeutic, offering the opportunity for addiction psychiatry to raise its low status in medicine and within psychiatry.

The Clinics had dropped their attempts to prevent the growth of a smuggled heroin market when they moved towards short-term oral methadone detoxification in the 1970s. The new policy favoured restricting the supply of tradeable pharmaceuticals and created an incompatible situation between different prescribing styles. While a doctor who prescribed liberally might have coexisted unproblematically with other services which only offered detoxification, one who considered that detoxification was the sole valid approach might view the existence of other services providing long-term prescription as undermining his or her work. This desire to standardise practice was one reason that the London consultant psychiatrists met regularly at the Home Office: 'Most of us took the view that we all needed to do much the same thing, so that people couldn't work their way round to find the most liberal prescriber.'[117] It also helped overcome their isolation and enabled them to share practical information. Thomas Bewley made clear that a contributing factor in ceasing injectable prescribing was because his colleagues were doing so, 'It would have been quite difficult for one consultant to prescribe in a markedly different way to the other units.'[118]

Thus Clinic services believed they needed to present a united front so that they *all* offered only short-term detoxification. Patients seeking treatment would then be forced down this path for their own benefit. If one service stepped out of line, patients would inevitably be attracted away by the offer of prescribed drugs, risking overdose, selling their surplus drugs, or deepening their dependence, pushing higher their dose and tolerance, making eventual detoxification more difficult. This implicit paternalism characterised drug users as unable to judge their own interest, not to be subjected to the 'temptation' of larger scripts.[119] This might have been more typical among psychiatrists, whose qualifications gave them legal powers to certify that certain patients did not know what was in their best interest and there might be greater potential for disagreement between patient and doctor on the diagnosis and appropriate treatment. Aside from public health in the control of infectious disease, psychiatry was the only medical specialty where a patient could be detained against his or her will for treatment (although not simply for addiction).

A treatment that provided choice regarding prescribing might instead have seen them as 'consumers', more compatible with the concept of a medical marketplace inhabited by private practitioners. Some doctors such as Dale Beckett, the addiction consultant at Cane Hill Hospital, who was outside the London psychiatric establishment and had worked in both the private sector and the NHS, questioned the very role of the state in controlling access to drugs.[120] Of course, a united front could have offered a more liberal rather than a more restrictive prescribing regime across the board but it is worth remembering that drug use was seen by some medical professionals as a moral issue arousing strong disapproval. Bewley had described his own misgivings about the 'sinfulness of pleasure' from drug use in an article in 1970.[121] His comment that 'we're not in the business of prescribing happiness drugs',[122] in explanation for his refusal to prescribe cocaine or amphetamines, might have also explained his strong preference for methadone over heroin in the early 1980s. James Willis, a dissenting London psychiatric consultant, attributed the move away from maintenance heroin prescription as partially due to doctors' tendency 'to moralise about their fellow creatures'.[123]

The tendency to standardise treatment across the Clinics towards the end of the 1970s was accompanied by an increased application of bureaucratic rules,[124,125] including contractual agreements between patients and staff regarding attendance and a number of other areas which had previously been subject to individual judgement.[126] The London psychiatrists' success in imposing bureaucratic rules on their Clinics could partly be attributed to the lack of counterbalancing forces.

Demands for services to be designed around the drug users' preferences rather than those of the providers were hardly heard within the Clinics at this time. Only sporadically since the 1980s have British drug users organised themselves to lobby for their interests in treatment. Some evidence has suggested that patient autonomy was actively resisted by doctors working in the field through the universal treatment model.[127] Patients' voices were weak because of their socially stigmatised status, poor collective organisation, desire for confidentiality and fear of losing their supply of prescription drugs. Clinic services were therefore not planned around the priorities of their users.

At the other extreme, private medicine, being more market-led, was more clearly influenced by patient preference. As very small organisations, private and general practices did not need bureaucratised systems and operated as individual businesses with non-standardised codes of behaviour. Some attempts were made to bring peer influence to bear among this disparate group in the 1980s by AIDA through the expulsion of members thought to be practising poorly, but it failed to get concerted support among private doctors and the organisation stopped meeting in 1988 (see Chapter 6).

Some have seen the proliferation of clinical guidelines as a symptom of decreasing medical autonomy and bureaucratisation resulting from employment of doctors by the state leading to a diminution of professional status.[128] While the Clinics had undoubtedly become increasingly bureaucratic, the use of bureaucratic rules actually served the leaders of the psychiatrists in their self-regulation to preserve and extend their prestige and their control over doctors outside the Clinics.

Concern about 'diverted' pharmaceuticals

The diversion of prescribed pharmaceuticals formed a major part of the argument in favour of controlling the prescribing of doctors outside the Clinics,[129] yet during the 1980s, this market was dwarfed by the large amount of trafficked heroin entering the country. Why then did this relatively small market prompt the range of measures proposed by the *Treatment and Rehabilitation* report, including the *Guidelines*? Was it simply ammunition used by doctors pressing for stricter prescribing against those not conforming to their treatment model or were there other reasons?

Unlike less tangible components of doctors' practice, prescribing was quantifiable and so more easily subject to criticism and standardisation. A visible market existed in diverted pharmaceuticals in London, to which attention was drawn by anthropologist Angela Burr in 1983.[130] This public revelation was seen as a threat to the perceived professionalism of doctors. In addition, there was evidence that major change could be achieved in reducing the supply of diverted pharmaceuticals, as had occurred with the amphetamine *Methedrine*, which, with the help of its manufacturers, had disappeared from the illicit drug scene by 1968.[131]

Conclusion

The origin of the *Guidelines* lay in the medical profession's claim to the sole right to determine treatment, as asserted by a psychiatrist member of the *Treatment and Rehabilitation* Working Group. In this doctors successfully defended their right to collective clinical autonomy against potential incursions from outside medicine. At the same time they were strengthening their collective clinical autonomy through the control, in the form of the *Guidelines,* of other doctors' individual clinical autonomy. Klein has described a similar situation after the 1991 NHS market reforms that witnessed the proliferation of clinical guidelines and protocols. Here the individual autonomy of NHS consultants was shrinking while they accepted greater collective responsibility.[132] However, in this case, the *Guidelines* were aimed not at regulating the addiction psychiatrists or the profession as a whole but the small number of private prescribers practising in and around London.[133,134] Significantly, four consultant psychiatrists and one NHS GP had served on the Working Group drafting *Treatment and Rehabilitation* but no private doctors.

The *Guidelines* were the result of a range of interested parties struggling to get their approach adopted as 'good practice', and appearing as a consensus statement from the profession. This was not unique in the formation of medical guidelines. Historian Jennifer Stanton's work on the development of the Hepatitis B vaccine policy guidelines showed how the epidemiology and potency of the disease played some part but were not the chief determinants of policy.[135] In the case of the treatment of drug users, not only was research evidence on the efficacy of treatments very limited at that time but there was also a lack of agreement within the profession on what drug dependence actually meant and over doctors' roles in relation to the drug supply.

The *Guidelines* showed the struggle for dominance of one treatment model – that of the London psychiatric establishment – against a range of interests represented in the Medical Working Group and its alliance with the bureaucratic interests of the state to achieve this. While publicly presented as a way of encouraging doctors to treat drug users, they were originally intended to be used for disciplining doctors, particularly private prescribers, who did not follow them and were later employed for this purpose.

The *Guidelines* were the codification of a change of practice achieved informally through peer pressure among the London Clinics, which could offer addiction psychiatry greater professional prestige and sense of achievement in their work. The Clinics' policies had not been driven by research on treatment effectiveness, but were justified retrospectively through the misrepresentation of one particular piece of research, the Hartnoll–Mitcheson trial. While this change of practice had been achieved informally through face-to-face contact among the London psychiatrists, it faced resistance and challenge from doctors outside, who, in turn, used little published evidence to justify their own positions. The *Guidelines* embodied the extension of this pressure towards prescribing conformity to doctors outside the Clinics, with the authority of medical 'consensus', and the threat of enforcement by the Home Office Drugs Inspectorate.

4

Ambiguous Justice: The General Medical Council and Dr Ann Dally

Introduction

In the 1980s, as co-founder and leader of the Association of Independent Doctors in Addiction (AIDA), Ann Dally spoke out vehemently against the Clinic system. During this time the General Medical Council (GMC) brought two disciplinary cases against her. The process of these cases shaped the role of AIDA in the public-private debate and revealed much about the GMC and the regulation of addiction treatment in the 1980s. Their outcomes affected private prescribers' leadership and their representation in the policy community for years to come. Commentators then and since have asked whether there was a connection between Dally's outspoken criticisms and the Council's actions.[1] Were they attempts to silence a dissenting voice or had Dr Dally's care fallen below a recognised standard? What do these cases tell us about medical self-regulation at this time?

Dr Dally's cases raised questions about the impartiality, consistency and transparency of the GMC.[2] They also brought up issues about the GMC's interplay with other regulatory bodies and the questions of whose interests were served by professional self-regulation and who decided who was fit to practice. Ann Dally was not the only addiction doctor to be taken before the GMC in the 1980s and her high profile made her somewhat atypical. Herman Peter Tarnesby, a lesser known private addiction psychiatrist, was subject to the same process and his case is examined here for comparison.

The General Medical Council

The origins of medicine's organisation and regulation as a profession date back to the 16th century but it was not until the 19th century that

Britain's doctors arranged themselves into bodies to represent themselves nationally in the form of the British Medical Association and, with state support, to regulate themselves through the General Medical Council. The 1858 Medical Act designated the GMC as the medical register to identify qualified doctors and gave the Council jurisdiction over professional conduct with powers similar to a legal tribunal.[3] Since then the degree of state involvement in the 'self-regulated' profession has varied and in the early 20th century government introduced special controls over doctors prescribing controlled drugs.

The freedom that allowed the medical profession to self-regulate was based upon the idea that only a doctor's peers were capable of judging performance. Over the last three decades of the 20th century doctors attempted to defend this principle while under increasing pressure. Patients, the academy, the media and later the government challenged the exclusivity of medical expertise. The GMC experienced major changes either imposed from outside or made under the threat of such intervention. From the 1960s cultural shifts across Western societies had started influencing a whole range of social and economic relationships, including those between doctors and patients. Civil rights movements challenged accepted social norms and the intellectual anti-psychiatry movement asserted that many mental 'illnesses' were socially constructed, questioning the basis of medical power. Inside the medical profession, the early 1970s saw a crisis precipitated by major dissatisfaction with the representativeness of the Council, new fees, and its treatment of overseas doctors. This led to a government inquiry, the Merrison Committee, culminating in the changes which among other things, divided disciplinary proceedings into 'professional conduct' and 'health', distinguishing the 'bad' from the 'mad'.[4]

From 1970, the GMC had taken the view that prescribing or supplying drugs of dependence 'other than in the course of *bona fide* treatment' constituted serious professional misconduct,[5] although it failed to define what 'bona fide' treatment was. Further, the Council seemed reluctant at this time to get involved in cases concerning errors in diagnosis, treatment or any issue that bordered on doctors' clinical autonomy,[6] including controlled drug prescribing. Its greater alacrity towards prosecuting doctors prescribing drugs of dependence in the 1980s may have reflected wider pressures on the Council to scrutinise its members' conduct more closely and openly.[7] Pushed into a defensive position, the GMC increased the number of cases it dealt with concerning conduct issues in the 1980s and 1990s,[8] providing an opportunity for those interests in the drugs field keen to exercise self-regulation for their own particular concerns. It also warded off the threat of externally developed legislation being passed by developing its own scheme to extend its own powers in disciplinary cases, eventually passed in legislation in 1995.[9]

Action and inaction

In the USA from 1919, prescribing substitute drugs to opiate addicts on a maintenance basis became illegal after passage of the 1914 Harrison Act and the Supreme Court's 1919 clarifications and from 1974 the practice was only allowed in very restricted forms. By contrast, British law has taken a limited role in controlling the way doctors treat drug users for addiction. Formal regulation of doctors' prescribing was largely left to the Home Office Drugs Inspectorate and to an initially unenthusiastic GMC. In 1967–68 the GMC had failed to act against private prescriber Dr John Petro who had aroused great interest among the tabloid newspapers by prescribing in London's underground stations and other public places. The Council had instead waited for the Courts to act, only erasing him from the register after the Home Secretary had withdrawn his powers to prescribe 'dangerous drugs'. Even then, a delay between the GMC's ruling to erase Petro from the register and his appeal hearing five months later, allowed the doctor to continue prescribing the amphetamine *Methedrine*. The Council had been 'greatly criticised' for the delay, and the loophole was closed by the 1969 Medical Act,[10,11] after which the GMC began to deal with cases of drug prescription a little more frequently.[12] Drug use and dependence also became more common around this point, rising from initially tiny numbers throughout the 1960s[13] but this was probably not the only reason. Public and political concern were heightened, which had in turn prompted the major changes in treatment services and legislation of the late 1960s,[14] establishing the Clinics and nurturing the new group of addiction psychiatrists.

Although the GMC had increased the number of controlled drug prescribing cases it dealt with, this did not reflect any greater enthusiasm for the issue. In 1971, Lord Cohen, then president of the GMC, spoke during the passage of the Misuse of Drugs Bill in the House of Lords, saying that without an extension of its jurisdiction by statute and increased financial support the Council could not investigate these cases, as 'we are not a police force; we have no inspectorate'. He urged the House to reintroduce the Home Office's Tribunal system, which it did in the passage of the Bill.[15] This provided a panel of doctors to judge whether one of their peers had been guilty of 'irresponsible prescribing'. Between 1973 and 1997, the Home Secretary could deprive a doctor of the right to prescribe controlled drugs but only with the agreement of other doctors. It seems likely that the objection of having no police force was a lobbying tactic to bring back Home Office Tribunals, rather than based in any desire to rectify this. In 1975, the Merrison Inquiry recommended that the GMC set up its own investigation unit to research allegations against doctors,[16] but the Council rejected the idea as inappropriate to its role.[17]

Despite this public criticism the Council's guidance on the treatment of drug dependence remained very limited through the 1970s and 1980s. Ann Dally reported writing to the GMC in August 1982 asking for advice on the treatment of addicts in private practice. She quoted the reply as stating 'the Council has hitherto issued no specific guidance' on that subject.[18] Indeed the GMC's guidance book *Professional Conduct and Fitness to Practice* only referred to 'the prescription or supply of drugs of dependence otherwise than in the course of bona fide treatment'.[19] From 1981 to 1985 the Council was considering issuing further advice on prescribing opioid drugs, particularly in private practice but seemed unable to reach a decision. When called to account by the House of Commons Social Services Committee in 1985, GMC representatives cited the Department of Health's 1984 good practice guidelines as sufficient to set out 'a corporate view of what constitutes proper practice in this field'.[20,21]

Between 1972 and 1984 the GMC's Professional Conduct Committee (PCC) had heard 39 cases of 'improper' prescribing and erased 18 doctors from the medical register.[22,23] Between 1983 and 1989 it greatly increased its prosecutions with 46 cases in six years.[24] These figures are difficult to interpret, however, since both tallies included doctors who had committed offences under the Misuse of Drugs Act as well as those whose manner of prescribing was considered problematic but not criminal. Furthermore, before 1980 these cases also included self-prescribing by addicted doctors. In the first years of the 21st century the GMC started to pursue private prescribers with an alacrity never seen before. It prosecuted seven doctors in the 'Stapleford' case, the largest GMC hearing to date, and by 2010 six of the eleven working private prescribers interviewed for this study had been erased from the medical register.

Relations between the GMC and the state

When dealing with drug dependence, the GMC had two points of contact with the state: the Department of Health, responsible for the NHS, and the Home Office Drugs Inspectorate. The relationship between the GMC and the state has been a complex one: although independent of direct government control, it was also 'part of the apparatus of the central state'.[25] The NHS would only employ doctors registered with the GMC, which was not technically a requirement in the private sector.[26] Throughout the period the Department of Health and Health Authorities as employers or contractors could exercise certain controls over NHS doctors but had no such powers over private doctors.

Within the GMC the dominance of NHS members could be perceived, as during the 1980s drug treatment expertise was recognised almost exclusively as residing within the NHS Clinics. The pharmaceutical profession displayed a mirror image of this relationship where the more numerous and better represented small business pharmacists dominated the salaried NHS employees in its professional bodies and in policy-making.[27] The Inspectorate, which is discussed in the next chapter, although using advice from NHS psychiatrists, tended to formulate its own independent views.

The Conservative governments from 1979 to 1996 and their relations with the medical profession revealed contradictory impulses within the British right wing. On the one hand, Margaret Thatcher's governments professed allegiance to the free market, clashing with the professions over their monopolistic practices.[28] Thatcher promulgated a radical social agenda that did not accept as given the privileged position of professionals and her stance tapped into a wider suspicion of hierarchy and deference. Along with the free market, came the exaltation of 'choice' and the supremacy of the consumer. From 1989 the government attempted to further extend consumer choice over the heads of the doctors with the introduction of an 'internal market' into the NHS. While other areas of the NHS saw a rise of consumerism, drug treatment remained resistant, while outside in the voluntary sector patient pressure groups and services could be interpreted as expanding consumer demand and choice, a role supported by state funding.

The 1997 New Labour government withdrew many of the previous government's market reforms but the rallying cries of consumer choice and medical accountability remained popular. Cooter has convincingly argued that the consumerist movement of the 1970s and 1980s broadened the base for participation in medical ethical thought, rhetoric and action, so that the turn of the 21st century saw the highest ever levels of claims for legal redress for unethical medical procedures and calls for statutory regulation to protect against unethical practices.[29] Government attempts at control became more overt in the late 1990s as several high profile 'scandals', such as the high patient death rate in paediatric cardiac surgery at Bristol Royal Infirmary, were cited as justification for state regulation. The GMC, under government, media and public scrutiny, wished to be seen to be doing its job and stepped up its activities considerably.

With the GMC responsible for all medical discipline including prescribing and the Home Office Inspectorate concerned specifically with controlled drugs, it was unclear during this period which body should take the lead. Despite Lord Cohen's declared distaste for dealing with this topic, ways of working seem to have developed between the Council and the Inspectorate without being made explicit by either side. As the Council itself had observed, the key difference between the bodies was that the GMC lacked inspectors to gather evidence for its hearings, relying on the Inspectorate for its information and to take the lead in Tribunals. Information flowed from the Inspectorate to the GMC but not in the other direction.[30] According to an inspector working since the early 1980s, the GMC 'saw us as in a sense doing their dirty work', a view Bing Spear reportedly shared at the time.[31,32]

While the Home Office automatically informed the GMC when a doctor had been convicted in criminal court, information about a Tribunal ruling would not necessarily be provided.[33] Reversal of GMC and Home Office sanctions were independent of each other, so that a doctor who had been erased or suspended from the medical register by the GMC and had their controlled drugs licence suspended by a Home Office Tribunal, could be re-registered by the GMC on an appeal or at the end of their suspension and would have to apply separately to the Home Office to regain their controlled drugs privileges. If working together, the GMC could re-register a doctor on the understanding that the Home Office would retain its ban on controlled drug prescription.[34]

The Inspectorate was an enthusiastic but discerning regulator throughout this period, using informal methods of persuasion in the early 1970s, while pressing for the return of its formal Tribunal machinery. The GMC supported these calls in order to relieve it of its own obligations, though still reluctantly prosecuting cases. Spear was gratified to see the Tribunal system reinstated but it was only used nine times between 1974 and 1982. In 1982 the Advisory Council on the Misuse of Drugs (ACMD) report *Treatment and Rehabilitation* had criticised the Home Office for underusing its Tribunal machinery, to which the Inspectorate responded by increasing Tribunals to four in 1984.[35] Unlike the GMC, who did not visit doctors in advance of initiating proceedings, the Inspectorate could use the threat of a Tribunal to influence practitioners, allowing it a more informal regulatory role. In the late

1990s the Council overtook the Home Office in both zeal and powers, as the Inspectorate once more lost its Tribunal machinery. For a few years the GMC continued to rely upon the Inspectorate for evidence against doctors. By 2009 the Inspectorate had changed its name and function and no longer inspected doctors, leaving the GMC to carry out its own investigations.

Ambiguity over the GMC and Inspectorate's roles could produce the strange situation of a double trial, as we will see with Dr Tarnesby who was taken before a Home Office Tribunal, his prescribing powers curbed, and then taken before the GMC to be tried on the same evidence on approximately the same charge. On occasion, as in Ann Dally's second GMC case, the Council took up cases that the Home Office had declined to put to Tribunal but the explanations for this were unclear. Much of the evidence given against Ann Dally in her GMC hearings was gathered by Home Office inspectors either through interviews or records of her prescribing patterns kept by pharmacies. The cases of Ann Dally and Herman Peter Tarnesby in the 1980s provide a window onto the period when the GMC and Home Office were both bringing disciplinary cases. Dally's cases in particular formed a turning point in the fortunes of private prescribers in England and a focus for the issues at stake.

Ann Dally and the GMC

Forceful, self-assured and articulate, Dr Ann Dally was the Oxford educated private doctor who started up AIDA in 1981 and became its first and only president. The 'Independent' in 'AIDA' referred to both private prescribers and NHS doctors working outside the Clinics.[36] Although claiming to seek closer co-operation with the Clinics it was directly oppositional in both membership and activities. Several AIDA documents opened with attacks on the Clinics.[37]

While not formally qualified as a psychiatrist, Dally had been working in private general psychiatric practice in partnership with her husband (and later, ex-husband), psychiatrist Peter Dally since the 1960s. By 1979 she was already known as a writer on medical matters and a respected doctor when she started treating opiate addicts in quantity. It was then that she formed her views that these patients were victims of the system of drug controls, forced into a criminal lifestyle to obtain their supplies. She attributed most of the adverse health effects of drug use to the illegal market rather than to the drugs themselves.[38] Ann Dally believed that long-term prescribing would allow addicts, who were unable to achieve abstinence, to live healthy, productive, law-abiding lives.[39]

AIDA's first meetings took place at the Home Office with Bing Spear attending.[40,41] He and Dr Dorothy Black, Senior Medical Officer at the DHSS, provided comments and contributions to the Association's draft clinical guidelines.[42,43] Dr Black was at pains not to 'take sides' in the dispute between doctors outside and inside the Clinics. In her response to AIDA's draft guidelines on clinical practice,[44] she disapproved of the document's criticism of the Clinics, chiding its authors, 'A responsible body such as your own should stand on your own practice rather than on a comparative exercise with that of others.'[45] So while civil servants concerned with drug policy were scrupulous in maintaining public distance and impartiality, Dr Dally was accepted and encouraged inside one part of the policy community as the respectable face of private practice.

Part of Dr Dally's intention in setting up AIDA was to raise standards among private doctors. In the early 1980s there was great concern over drug users taking *Diconal*[46] particularly as some were crushing up the oral tablets and injecting them with disastrous consequences. The drug was often obtained from doctors unaware of or indifferent to the way it was being used. In September 1982 AIDA resolved that 'the use of *Diconal*, except in the most exceptional circumstances, is incompatible with membership of our Association',[47] a declaration that was to trip up Dr Dally later on.[48] AIDA took its own regulatory action in 1983, expelling one member, Dr Rai, for prescribing *Diconal*.[49,50] Rai was then disciplined by the GMC the following year.

Ann Dally and Margaret Thatcher had been contemporaries at Somerville College, Oxford. In 1983 before the GMC initiated its case against her, Ann Dally visited Mrs Thatcher at 10 Downing Street, to express her views on drugs policy and treatment and to criticise the Clinics.[51] As a good networker, Dr Dally was successful in achieving access to policy circles but with little direct influence. Although apparently impressed by her sincerity, Mrs Thatcher did not take sides, and after the GMC's verdict in her first case, the Prime Minister wrote a reply to a letter from Dr Dally maintaining this line: 'I hope you will forgive me if I do not say anything about the circumstances of your case. But I know that this must be a painful situation for both you and your husband. I know too that the strength you have always shown will carry you through this difficult time.'[52]

In the three years preceding Ann Dally's first case, it was clear that the outward peace between the Clinics and the private prescribers had been broken and hostilities were polarising the field. Attacks came from both sides, through official channels such as the ACMD and in the media. Articles critical of individual private doctors in the tabloid press had appeared years earlier, with for instance, the *Daily Mail* and *The Sun* pursuing Dr John Petro in 1967.[53] However, it was not until 1980 that the first attack on private prescribers, by Thomas Bewley, appeared in the medical press, followed by further criticism by *The Lancet* medical journal.[54,55]

The ACMD's *Treatment and Rehabilitation* report included a strong attack on private prescribers and recommended a range of controls to regulate them (see Chapter 2).[56] *The Lancet* returned to the subject in March and April 1983, with Hamid Ghodse, a junior psychiatric colleague of Dr Bewley, defending the Clinics and attacking outside doctors.[57] This prompted Dr Dally to write forcefully to contradict him.[58] Bewley and Ghodse then teamed up together in what was perhaps the most significant attack in a medical journal, due to its use of 'evidence', its timing and its uncompromising title, the 'Unacceptable face of private practice: prescription of controlled drugs to addicts'.[59]

This uninhibited assault on private prescribing was published in the *British Medical Journal* only three weeks before Dr Dally's first GMC case. Dr Bewley avers that it was written before Dr Dally's case came to light, and that he was unaware of the timing. It was accepted for publication on 8th April 1983, before Dally's first GMC hearing, and prior to the decision on 12th May 1983 to take the case to a disciplinary hearing.[60] However, Dr Bewley did concede that the questionnaire was 'not a piece of serious scientific research' but had just been carried out to make a point.[61] The journal clearly wished to stir up controversy around private prescribing, featuring it, like Bewley's previous article, under the banner 'For Debate' and adding its own unsigned leader criticising both sides.[62]

The questionnaire on which the article was based was methodologically weak in its construction and the 69 per cent response rate Bewley himself later described as 'completely useless'.[63] One question asked drug users attending two NHS Clinics about the reasons why drug users attended private practitioners, but only half of the respondents had attended a private practitioner. Despite this, responses from all respondents were counted as valid, so that NHS patients were being asked to speculate as to the reasons for attending a private practitioner, including leading questions such as whether such doctors were 'more easily conned than Clinic doctors'. The article claimed that data from the Home Office Addicts Index showed a change in the previous three years so that 'large numbers of addicts' were having drugs prescribed for them by private GP. In fact, it was not possible to distinguish from the Home Office data used in the article whether the GPs were NHS or private. No quantitative data were collated for the research regarding the numbers of patients attending private practitioners or private prescriptions issued.

The article also claimed that the reason for 'such large numbers of addicts attending private practitioners' was that they prescribed *Diconal* and *Ritalin*,[64] the two drugs at the centre of Ann Dally's trial. It blamed 'uncontrolled prescribing by private practitioners' in the 1960s for 'a severe spread of addiction', despite the fact that only one of the doctors considered the source of this in the 1965 second Brain Committee report was working privately (see Chapter 1).[65-67] The article asked 'whether it was ever desirable to prescribe controlled drugs to an addict when a fee is paid'.[68] Bewley and Ghodse described 'an urgent need to control prescribing' of methadone, *Diconal* and *Ritalin*, either through the General Medical Council, the Home Office Tribunal system or an extension of the licensing system to include all controlled drugs. Bing Spear described the article as 'an authoritative establishment attack on the private sector' that 'presented a wholly false picture of the conditions prevailing in the generality of [Clinics]'.[69]

The *British Medical Journal*'s leader article in the same issue accused some private doctors of effectively selling drugs to addicts, who in turn funded their treatment by re-selling some of their prescriptions on the black market, but it did not spare the Clinics, which 'seem to have faded into decline'. It questioned their move to oral methadone, and called for 'new policy objectives' to contain the 'epidemic of drug use'.[70] Debate was unleashed and eight letters appeared in subsequent editions of the journal, both critical and supportive of Bewley and Ghodse's article. They ranged across the spectrum of the drug treatment community, including a (private) patient which was unusual for the policy debate at this time.[71] The array of responses covered most of the points which were to constitute the public-private debate over the 1980s and 1990s: the impact of substitute prescribing on the incidence of addiction; centralisation versus decentralisation of prescribing decisions; the sources of fees paid by patients; leakage from prescriptions to other users; the potential incomes of private prescribers; the role of the black market in trafficked drugs and the healthcare worker–patient relationship.

In addition to these printed words, Dally claimed that AIDA's Home Office meetings were forced to move to her own premises in Devonshire Place after the Inspectorate was pressurised by Bewley and Connell[72] as, 'Meetings there had given us a respectability that was unacceptable in some quarters.'[73] This coercion has been difficult to confirm, but Spear commented that, 'it was quite obvious the London consultants did not take too kindly to the contact the Drugs Inspectorate had with AIDA'.[74]

In 1983, Dr Dally was charged with issuing a patient, one Brian Sigsworth, with prescriptions for *Diconal* and methylphenidate (*Ritalin*) otherwise than in the course of bona fide treatment.[75] At the end of the hearing, the PCC took the view that Ann Dally had disregarded her special responsibilities as a doctor by prescribing drugs of addiction and dependence in large quantities; having taken insufficient steps to establish that there were adequate therapeutic reasons for doing so and for failing adequately to monitor the patient's progress and the use to which the drugs were being put. She was judged guilty of serious professional misconduct and admonished. Because Ann Dally was not suspended or erased from the register, she was unable to appeal against the verdict.

This case concerned in particular the prescription of the oral tablet *Diconal*, the injected use of which had become notorious in the preceding years. It was clear, however, that Dally's patient was not injecting the drug and the Council failed to trace back to Dr Dally the *Diconal* Sigsworth had sold. In 1981, when Ann Dally's prescribing occurred, there were no official guidelines on the treatment of addiction, no legal rules on specific matters such as dose, and the guidance given by the British National Formulary on *Diconal* related only to the treatment of pain and terminal disease.[76] Ann Dally was also criticised for not taking urine tests to check on her patient's consumption of the prescribed *Diconal*. Her defence argued that such tests were easily falsified by patients. Critically, and in an apparent extension of the GMC's definition of a doctor's duty, Dr Dally was considered responsible for the fate of drugs prescribed. She had only prescribed *Diconal* to five patients,[77] and had discussed the dose she was going to prescribe with a Home Office drugs inspector, Mr Heaton, although he was not medically qualified and the decision remained her responsibility.[78] The question of serious professional misconduct therefore seemed to turn upon the extent to which a doctor could be held responsible for the ultimate fate of the drugs she prescribed and to what extent she could be expected to predict this.

Between 1983 and 1986 the landscape of drugs policy changed in important ways. The prevention of HIV/AIDS emerged in public debate and the government started to sponsor needle exchange schemes. A great deal of discussion regarding appropriate prescribing appeared in the media, which was becoming more sympathetic towards long-term prescribing.[79] Not long after the first case, Dr Dally had felt apprehensive that a second was brewing. She had received a visit from two Home Office inspectors who warned her that the Clinic doctors or 'drug dependency establishment' were trying to get a Tribunal brought against her.[80,81] She wrote to Mrs Thatcher, saying 'I believe my views are shared by an increasing number of interested and informed people. Perhaps partly because of this I have aroused much hostility in powerful circles. I believe that I am again in danger of being "fixed" as happened last year.'[82] The Prime Minister was sympathetic in her reply but did not refer to the GMC issues and again refused to take sides in the dispute.[83]

After interviews, a report and some correspondence the Home Office took no action against Dr Dally but the GMC decided to use the evidence the Home Office inspectors had gathered to put forward its own case. The Inspectorate's decision may have been influenced by Bing Spear who generally supported Dally's work.[84,85] While taking action against some of the prescribers he considered less responsible, Spear seems to have recommended Dr Dally to at least one patient.[86] The GMC may have been influenced in the opposite direction by Philip Connell, one of the most senior London Clinic psychiatrists, an active Council member representing the Royal College of Psychiatrists, and strong opponent of private prescribing.

In September 1986, the GMC accused Dally of professional misconduct for a second time on two charges dating back to 1982: irresponsibly prescribing numerous controlled drugs in return for fees and irresponsible prescribing in return for fees in relation to a particular patient. The latter included an alleged failure to conduct a conscientious and sufficient physical examination, inadequately monitoring his progress on each occasion when a further prescription had been issued, and discharging him without making arrangements for him to receive ongoing care and treatment from another doctor. After a gruelling eight day hearing, the Council found Dr Dally guilty of serious professional misconduct in relation to the specific charge about Mr A but not in relation to the general allegation of irresponsible prescribing. She unsuccessfully appealed the verdict and was suspended from prescribing controlled drugs for the treatment of addiction for 14 months.[87] The Council failed to prove the general charge of irresponsible prescribing, and the appeal conceded that medical opinion was divided on the issue of long-term prescribing of controlled drugs to addicts.[88] In this Dally may have been assisted by British policy responses to HIV which had begun to strengthen the position of those advocating maintenance or long-term prescribing (see Chapter 1).

The patient had admitted to selling methadone ampoules prescribed by Dr Dally and the police had also proven this.[89,90] One of the accusations was that Dr Dally had failed to provide a referral after discharging him as a patient. However, the patient had turned up late and was afterwards abusive. Furthermore, the patient went to his GP two days later and got a referral to Hackney Hospital Drug Dependency Unit but decided not to take it up. To consider this 'serious professional misconduct' seemed harsh, particularly as her practice was exonerated of the general allegations in the first part of the charge.

At least one commentator has characterised Dr Dally's second trial as an inappropriate attempt by the GMC to adjudicate over different schools of thought of medical practice, namely long-term versus short-term prescribing, when agreement or even relatively stable opinion were lacking in the field.[91] There was much discussion of the appropriateness of long-term prescribing during the hearing but the fact that Dr Dally was cleared of the general charge of irresponsible prescribing partly vindicated her approach. The second charge was proven, most of the issues in it were matters of fact but whether they were serious enough to justify a disciplinary hearing and could reasonable be considered 'serious professional misconduct' by the standards of the day was questionable.

Following on from Ann Dally's 14-month suspension from prescribing controlled drugs the previous year the Chairman of the PCC judged that Dr Dally had failed to comply with the condition that was imposed on her registration as she had prescribed substances which were controlled under the Misuse of Drugs Regulations 1985 and subsidiary regulations. These were *DF118s* (which included dihydrocodeine), *Dalmane* (flurazepam), *Rohypnol* (flunitrazepam) and *Valium* (diazepam). However, no further penalties were imposed due to confusion over which drugs were covered by the term 'controlled'. The chairman concluded 'I have been asked to make it clear that the committee regard the term "controlled drugs" in that condition as meaning all drugs which are specified in Schedules 1–5 of the Misuse of Drugs Regulations 1985.'[92] She regained her full registration and ability to prescribe controlled drugs on 14th November 1988 but by then had retired from practice.

The conflicting nature of the advice given to Dr Dally by the GMC and various official sources as to which drugs were 'controlled' was attested by one of the Home Office inspectors involved.[93] The British National Formulary and similar prescribing handbooks only marked with a 'CD' denoting 'controlled drug' those in Schedules 1–3, which were also the only ones subject to requirements for prescriptions to be handwritten, leading one commentator to remark, 'This case demonstrates nicely the great care and precision which is required in imposing conditions, and the desirability of explaining precisely what is intended to the practitioner.'[94]

Ann Dally has alleged that the drug dependency 'establishment' made up of psychiatrists working in the London Clinics led by Connell and Bewley, were instrumental in the two GMC cases against her,[95] their intention to silence or discredit criticism. She had been warned in April 1984 that this establishment was 'still trying to make trouble' for her and were trying to have her charged before a Home Office Tribunal. According to Dally, and one of the inspectors present at the meeting, she was advised, 'You will be judged by the standards of the Clinics and if found wanting you will be deprived of your right to prescribe controlled drugs. It will all depend on how much you conform to what the Clinic doctors want.'[96,97]

While it has been difficult to trace the behind-the-scenes activities and complaints that led to the GMC cases, there were some pieces of evidence that were suggestive. The first case seemed to support Dr Dally's argument of malicious intent towards her as it concerned a fairly trivial matter: a single patient who had sold some *Diconal* which may or may not have been prescribed by Dally. Although there was considerable concern at the time that *Diconal* was being injected with dangerous results, even within AIDA,[98] it was clear from the case that the patient in question had not injected it.

Home Office inspectors confirmed that extensive checks had been made by Dr Dally with the Home Office Drugs Branch regarding the patient at the centre of the first case when she agreed to take on his care.[99] She had obtained information about his criminal record, finding that he had no records for supplying controlled drugs, and discussed the dose of *Diconal* that she was intending to prescribe. Attempts to get Dr Dally taken to a Home Office Tribunal may have failed and a medically led body, of which one of her critics, Dr Connell, was a member, was used instead. Favourable testimony was given by the Inspectorate about Dally, although Bing Spear did say that he did not remember so high a *Diconal* dosage as she had prescribed.[100]

However, considering Dr Dally was aware that she was under scrutiny, she may not have helped herself in the subsequent years before the second case for which the evidence was a little stronger. This time the police did prove that Dally's patient was supplying drugs *she* had prescribed, after secretly marking ampoules dispensed to him. Although difficult to predict or prevent this, she ignored evidence that at least one of her patients was unemployed and so considered by regulatory authorities at risk of selling on part of his prescription. She had also discharged a patient, albeit one who had been abusive towards her, without arranging any follow-on care and had carried out minimal examination of a patient before prescribing to him, although he had come to no harm.

Going against the 'conspiracy' interpretation was Don McIntosh, a senior Home Office inspector who acted in Spear's place during his frequent sickness absences in 1985. He was *not* part of the 'drug dependency establishment', but rather one of a range of voices within the Inspectorate. Coming from the Bradford Office in the North of England, where private prescribing was virtually unknown, he felt that different standards were being allowed in the South East in terms of the quantities and range of drugs prescribed to addicts. On moving to the London office of the Inspectorate in 1984 or 1985, he stepped up interviews of private doctors and in his report on Ann Dally recommended a Tribunal.[101] However, Peter Spurgeon, Spear's immediate successor, has suggested a contrary view that McIntosh may have been reflecting pressure from the Clinics that Spear had been able to resist.[102]

Dr Dally has argued extensively that the GMC was unfair in its conduct of the cases against her, believing it showed favouritism to its own members, vindictiveness and inconsistency.[103] One of her points was supported by Dr Michael O'Donnell, a member of the GMC's PCC, who argued that the committee members were allowing themselves exemption from their own ethical guidelines by allowing information from patients' notes to be used without their permission (provided the patients were not named) in Dr Dally's second case, and he withdrew from the case in protest.[104]

A memorandum submitted by the GMC to the House of Commons Social Services Committee on 20 February 1985 suggested that the Council had taken its own line on appropriate treatment for drug users prior to this case. It read:

> The Council has hitherto eschewed the promulgation of specific views on the correct regime of treatment for a particular condition: if the Council promulgated such views it would tend to inhibit advances in therapeutics. Nevertheless, disciplinary inquiries into cases of this kind have all too plainly demonstrated the special hazards of medical practice in the field of prescribing to addicts, particularly when a doctor is in practice on his own. The prescribing of opioid drugs to addicts, unless it is strictly controlled by the practitioner, may foment the growing problem of drug abuse, by increasing supplies of the illicit drug markets, rather than achieve the therapeutic aims of control, alleviation and detoxification. In the public interest, the Committees have felt bound to take a grave view of cases

where it was proved that a doctor had undertaken such prescribing irresponsibly or otherwise than in good faith.[105]

A clear injustice against Dr Dally could be seen in the PCC's final judgment delivered by the chairman, who restricted Dr Dally's prescribing, in the light of her 'blatant failure to heed the warning conveyed' by her 'previous appearance before this committee in 1983 in relation to similar matters', since part of the charge proven in the second case – the inadequate examination of her patient 'Mr A' – occurred in 1982 before her first hearing.[106]

Whether Dr Dally's membership of the group that produced the 1984 clinical *Guidelines*[107] counted against her in these cases is also hard to determine as they were quoted by both Counsels in their arguments. The prosecution referred extensively to their warnings against long-term prescribing, particularly of opioids, without specialist collaboration (i.e. from the Clinics).[108] But when a consultant psychiatrist from a Drug Dependency Unit in Brighton gave evidence for the Council, Dr Dally's defence compared his Clinic's prescribing and showed that some of his patients received maintenance prescriptions against the *Guidelines'* advice.[109] When interviewing Dr Dally with fellow inspector John Gerrard, Don McIntosh asked whether she disagreed with the *Guidelines* but he conceded under cross-examination that they were only advisory and a doctor favouring a different treatment regime would not necessarily be acting irresponsibly.[110]

If Dr Dally's opponents wanted her GMC cases to make an example of poor practice among private prescribers, the weakness of the charges and evidence against her made her a bad choice. Dr Tarnesby, whom the GMC erased from the medical register in 1984, would have made a much more dramatic example. If, as Dr Dally claimed, they wished to drive her from the field, then the second GMC case was successful, but although this is probable, it is still unproven. With the departure of Dr Dally from the scene, private prescribers lost their strongest representative. Without her leadership, AIDA withered away and the private prescribers lacked representation until 1996 (see Chapter 6). In a sense, Dr Colin Brewer, founder of the Stapleford Centre, a private drug and alcohol clinic in London, inherited Dr Dally's mantle. He was a member of AIDA and like Dr Dally saw prescribing as a broader political issue touching major social questions. He too wrote on medical matters in the press and saw addicts as victims of an overly restrictive regulatory system for controlling the availability of drugs. When Dr Dally ceased her practice after the second GMC case, he took on many of her patients.

Ironically, in 2004 he and his practice became the subject of the largest GMC disciplinary hearing of private doctors ever held and he was struck off the medical register in 2006.

Dr Herman Peter Tarnesby, second case, 1984

The story of the 1984 *Guidelines* showed how a mechanism for maintaining and raising standards of care and identifying cases of poor practice was hijacked by one ideology to dominate another. Some of the same tendencies could be seen in the Dally cases, but was this the case for all the Council's discipline against private prescribers over this period? A detailed review of every case has not been within the scope of this study but a contrasting case study of Dr Tarnesby suggested that in its dealings with private doctors the Council also played some role in protecting patients from incompetent or negligent practitioners. It is unclear whether this was enough to actually safeguard patients or merely provided a veneer of activity to protect the profession's claims to self-regulation.

Dr Tarnesby was highly qualified, with a doctorate in psychological medicine and extensive experience as a psychiatrist. As a contemporary of Dr Bewley's they had both trained at the Maudsley Hospital in the 1950s. Tarnesby went on to the respected Tavistock Centre (1952–59). He had been appointed consultant psychiatrist at the British Hospital for Functional Nervous Disorders and had worked with some drug-dependent patients as a consultant at Paddington Hospital, although it was not clear whether this had involved any prescribing. Dr Tarnesby then worked as a private psychiatrist, with consulting rooms in and around Harley Street, with only a little contact with drug users until he started treating them in quantity from 1981.[111–112]

His first brush with the GMC had occurred in 1969 when he was found guilty of serious professional misconduct for advertising abortion services.[113] The GMC charged him with prescribing both irresponsibly and otherwise than in the course of bona fide treatment in 1984. Since a Home Office Tribunal had already proved him guilty of irresponsible prescribing the previous year, he only contested the accusation of non-bona fide treatment.[114] Although Dr Tarnesby went to some lengths to research and refine his treatments for drug users, even commissioning the production of special methadone suppositories to avoid the need to prescribe injectables, he also seems to have made some serious errors of procedure and judgment.[115] He prescribed drugs to a patient whom he had not examined thoroughly and turned out later to be an undercover reporter for the *Daily Mirror*, treated several patients who subsequently died of overdoses using drugs he had prescribed, and kept inadequate records.[116–118]

There were a number of similarities with Dr Dally's cases, which have pointed up the difficult position drug doctors could be put in by the regulatory authorities, such as whether to discharge a patient who was not meeting their fees for fear that they could be selling some of their script. Also, the practice of the Clinics seems to have been taken as the ideal against which other treatment had to be measured, reflecting the stronger position of the Clinics within establishment bodies such as the GMC.[119,120] Overall, the evidence did show a carelessness that turned out to have serious and even fatal consequences for his patients.

Defining terms

The lack of definitions in the GMC's code of practice regarding 'bona fide' and 'irresponsible prescribing', the latter term also undefined in its inclusion in the Misuse of Drugs Act 1971, had left much latitude to doctors' clinical judgment. But this freedom could also be a trap as it allowed regulators, whether the state or professional peers, equal scope to interpret these terms as they chose. Clarification could be brought by the Legal Assessor, a lawyer advising the PCC, as was the case with the final definition of 'bona fide' used in the first Dally case.[121] However, the Legal Assessor's definition did not quell concern among commentators. Diana Brahams, a barrister writing for *The Lancet* after Dally's admonition, considered 'disquieting' the way in which the charge of prescribing drugs 'otherwise than in the course of bona fide treatment' was interpreted by the PCC.[122]

Brahams was concerned that the term was only defined as 'recklessness' at a late stage of the proceedings but then this was found to be unsuitable. Definitions were then provided for 'bona fide' which seem to have amounted to recklessness, making the ruling inconsistent. If the term meant, in literal translation, 'good faith', Brahams further argued that the evidence against Dr Dally seemed 'to fall well short of proof of a lack of good faith'. Certainly considerable care seems to have been taken by Dr Dally to prevent the prescribed *Diconal* from falling into unintended hands and Ms Brahams concluded her criticism of the GMC by calling for 'more positive guidelines and procedures ... for the private management of drug dependence'.[123]

Confusion continued in subsequent cases. In the Tarnesby case the following year, the defence spent considerable time trying to define the vaguely written charges, including the problematic 'bona fide'. The Legal Assessor stepped in again, not as might be expected to refer back to the earlier definition in the Dally case, but simply to translate the term into 'good faith'.[124] Dr Tarnesby's defence also had difficulty over whether the charge, which referred to the quantity of drugs prescribed meant *overall* or *per patient*. The time period covered by the charges was ambiguous too as was the phrase 'prescribing in return for fees',[125] used also in Dr Dally's second case. Since private doctors charged fees and provided prescriptions during the course of their consultations, it would be difficult to distinguish clearly when a fee was being charged for a prescription and when for a consultation, weighting the system against private prescribers. Even the term 'controlled drugs' in Dr Dally's second case was not satisfactorily pinned down until too late.[126]

Although the GMC had failed to advise its members on how they should prescribe to drug users and avoid regulatory attention, after 1984, as a spokesperson explained to the House of Commons Social Services Committee in 1985, there were other sources of guidance. By the time of Dr Dally's second case, doctors working privately had, according to the GMC's prosecution, four key sources of written advice: the 1984 *Guidelines*, the passing reference in the GMC's 'Blue Book', and two articles by London Clinic psychiatrists in the *British Medical Journal*. However, none of these were based on research evidence and like Dr Dally's practice and beliefs, they were effectively the product of personal experience and opinion.[127–130]

In the Tarnesby case the role of witnesses pointed up the problems around 'expertise' in this polarised, politicised field, and the potential conflict this could produce within a system of regulation based upon professional consensus. Dr Bewley, a vocal opponent of private prescribing, was called as a witness for the Council. Dr Tarnesby had prescribed to one of Bewley's long-standing patients, causing Dr Bewley to write him a vigorous letter of complaint. A second patient of Bewley's who went to Dr Tarnesby for treatment died of a *Diconal* overdose and Bewley had given evidence against Tarnesby at his Home Office Tribunal the previous year. In spite of Dr Bewley's clearly opposing position, he was treated as a neutral 'expert' by the committee, who saved a question of pharmacology arising earlier in the proceedings for him to answer.[131]

Uncertainty also characterised the nature of the GMC's disciplinary powers. It had the legal powers of a tribunal and required the level of proof to be the same as a criminal court, 'beyond reasonable doubt'. However, the charges could be specific or very general and unattached to any particular patient. The dates to which these charges applied could also float freely. Dr Tarnesby's charge was situated 'Between about 13 October 1981 or earlier and about 10 February 1983 or later...'[132] The transcripts of these hearings give the impression that the committee members themselves were unsure of their roles, perhaps unsurprising in view of the minimal preparation they were given.[133] Legal counsels too might be inexperienced in the ways of the GMC: Dr Tarnesby's defence was unused to the niceties of medical confidentiality, repeatedly revealing the identities of patients through the proceedings.

The media

The media acted as both a conduit for the views of either side of the debate and as an actor in its own right. There was an important contrast in the way that Ann Dally and the London consultants used the media, which may have had implications for the actions taken against her. The consultants published articles and letters in the medical media,[134] but very rarely took the debate to a general audience through press, television or radio. Already an established medical commentator, Ann Dally was prolifically outspoken and in the 1980s began to write many letters to the general press and appeared on the radio and television.

The tradition of treatment policy-making in a closed world of committees between experts and civil servants[135] rarely involved patients or public debate. It is the conclusion of this research that it was the public nature of Ann Dally's attacks on the Clinics, as much as the content of the attacks themselves, that so embittered the London consultants. Raising issues in public broke the consultants' code of discreet, private policy-making and involved the public and patients, whom they would have preferred to exclude, in the issues. Dr Dally raised policy questions with the ordinary public and openly criticised the Clinics while her key opponents, Drs Connell and Bewley, almost always restricted their opinions to medical fora, such as the *British Medical Journal* and *The Lancet*. Publicity did not always serve Dally well, however. Although she received encouragement from William Deedes, editor of the *Daily Telegraph*, in the early 1980s journalists in the general press and particularly the tabloids were often hostile.[136] Widely featured in the media, Dally often felt misrepresented.[137]

As well as the medical press and general media criticising both the Clinics and private doctors,[138] the tabloid press took a more active approach using undercover reporters to pose as drug-dependent patients to test the ease with which they could obtain prescriptions from private prescribers. In the case of Dr Tarnesby the resulting article in the *Daily*

Mirror prompted investigations by the Home Office Drugs Inspectorate and were also heavily featured in disciplinary cases before the GMC.[139] The article by Bewley and Ghodse and the correspondence that followed was provided as background material to the Medical Working Group responsible for producing the 1984 *Guidelines*,[140] which played a role in Ann Dally's second case. Around this time Dr Dally was also participating in a Thames Television programme 'Reporting London' on the prescription of *Diconal*.[141] Private was 'public' and public was 'private'.

Following the verdict of the first case, Dr Dally received sympathetic letters and coverage from journalists at *World Medicine* and *The Lancet*. Penny Chorlton of *The Guardian* wrote to her, 'I do feel you were made a scapegoat for challenging the establishment's approach [sic] to drug addiction'[142] and GMC member Michael O'Donnell also wrote in the *British Medical Journal* against the verdict.[143] The second GMC case against Ann Dally aroused much more attention both from the public and in the medical world as the prevention of HIV/AIDS became an important discussion. The prosecution feared that the publicity would 'turn the inquiry into a political debate'.[144]

Mike Ashton, editor of the Institute for the Study of Drug Dependence's trade journal *Druglink*, characterised the two Dally cases as political in origin: 'The powerful tide of medical opinion that wants prescribing more tightly controlled' was extending the GMC's powers from assessing treatment of the individual patient to the question of whether any drugs of dependence prescribed might be redistributed and harm other members of the public.[145,146] If they had wanted to silence her, the media attention that both Dr Dally's GMC cases drew rather backfired on her detractors. Press and public were able to sit in on the hearings and they and Dally drew the debate into the public realm, beyond the medical media, widening it to include the GMC process itself and the justice of its decision.

Doctors not only had their own publications, which also had a standing outside of their professional circles, but also easy access to the non-medical media. There was considerable public interest in medical issues throughout this period and as a medical professional, a writer or broadcaster held an automatic authority. For the tabloids, the shock value in undermining an apparently respectable figure by duping him or her in an undercover operation was all the greater. An emotive topic, drugs divided the public as much as the profession and ensured a readership. All aspects of this output fed the public-private debate not only at the rhetorical level but in its expression through regulatory action, whether in the form of guidelines, through the Inspectorate or the GMC.

Conclusion

The role of the GMC was problematic in the Dally cases for a number of reasons. A key plank in medicine's self-regulation was the idea of professional consensus, something clearly lacking in the drugs field during the 1980s, and to a slightly lesser extent in the adjoining decades. The lack of guidance as to what could lead to disciplinary steps and even on the conditions imposed on prescribing after a verdict, created an unfair situation for doctors and allowed scope for redefining ambiguous terms to suit personal or professional animosities. Ambiguity pervaded prescribing regulation in England throughout the jurisdiction of different regulatory bodies and in the guidance given to doctors about prescribing to drug users. Baker has traced back British medical ethics to a code of honour of the 18th and 19th centuries, where, by virtue of being a gentleman, a doctor was not deemed to require precise, codified guidance. Indeed the need for such explicit instruction on ethics and conduct could mark one out as unsuited to practising medicine.[147] Gentlemanly status, however, was something most doctors aspired to rather than achieved at this time. Particularly before the 1858 Medical Act and the establishment of the exclusive Medical Register, most ordinary doctors in England were of low social status.[148]

After the Second World War the British medical profession developed a more codified set of medical ethics but resistance to explicit advice continued. The Merrison Inquiry rejected the idea of a code of practice to give doctors a better idea of what might lead to disciplinary actions in favour of building up 'case law' as had been done in the past.[149] But 'case law' was used inconsistently, with rulings at one hearing not carried over to subsequent ones, and the same confusions arising repeatedly. The Council's reluctance to be pinned down in giving guidance to its members could also be seen in its laggardness to decide on whether to expand its advice on prescribing opioids. Its repeated unwillingness to state definitively what constituted non-bona fide or irresponsible prescribing, the lack of guidance from either the British Medical Association or GMC on the treatment of drug users and the uncertain meaning of the Misuse of Drugs Act term 'irresponsible prescribing' reflected both the profession's discomfort at judging the clinical decisions of other doctors and the uncertainty of the drugs field itself in the 1960s, 1970s and early 1980s. Competing schools of thought with different treatment goals, the lack of a robust scientific evidence base, and a relatively low level of technical expertise required for treating drug users, all made competence difficult to define. During the 1980s this vagueness, particularly when the GMC was increasingly being called to be specific, led to a situation exploitable by forces keen to restrict prescribing particularly among those in private practice.

The profession had a poor record of concern for regulating the treatment of patients, particularly when these patients were socio-economically disadvantaged.[150] During the 1970s and 1980s, the rise of patients' rights and consumerism outside of the profession increasingly pressurised the GMC to address issues of clinical decision-making, especially when it involved neglect, harm or death caused by practitioners.[151] Stacey noticed a rise in disciplinary cases concerning doctors' conduct in the 1980s and 1990s and the case of Dr Tarnesby showed that the GMC did fulfil some role in protecting drug-using patients from private prescribers whose practice was dangerous, however reluctantly.

Whether there was a conspiracy to remove Dr Dally as a thorn in the side of the drug dependency establishment has been difficult to prove for certain but some of the evidence pointed in that direction. Bing Spear, although not impartial, seemed convinced this was the case. Dally's first case was brought on a slim pretext and the procedure itself was flawed. The second case, though a little stronger, was still not damning, and she was doubly condemned for failing to heed the warning of the first, when some of its charges pre-dated it. Dr Dally was exonerated of the second case's general charge of irresponsible prescribing, which pointed against the idea that she was condemned for following a different 'school of thought'. Yet the fact that the minor misdemeanours proven in the second part of the charge were defined as 'serious professional misconduct' and brought suspension of her prescribing rights, has suggested a bias against her. Although Dr Dally was effectively driven out of the prescribing field by the two GMC cases, the media attention from outside medicine that they brought to the debate and to the Council's treatment of her rebounded on her critics. If part of their irritation was her high profile as a critic of the Clinics, the cases only brought her more publicity and some of the sympathy she received from the media was at the expense of her opponents.

While much attention has been given to the question of whether Dr Dally was being judged by the standards of the Clinics, no one has asked how much the GMC had absorbed the interests of the state in the form of the Inspectorate. Dally and Tarnesby's cases have shown that the Inspectorate's responsibility to control the flow of prescribed drugs within authorised channels had effectively been incorporated into the body of medical ethics for professional self-regulation. In their practice, private doctors were expected to distinguish between patients likely to divert drugs, to prevent their prescribed drugs reaching the hands of others and to keep monitoring their employment status, with limited means beyond the word of patients themselves and their own suspicions. Although advised by the Inspectorate, doctors were responsible for their own prescribing decisions, and the priorities of the state in controlling the circulation of drugs would not necessarily concur with the therapeutic or practical needs of the patient, for which they were also answerable. Ethical decisions about whether to treat had to take into account the patient's ability to pay from legitimate sources of income. Here, private doctors' practice had developed a criminal policing role that was not imposed upon Clinic doctors, while leakage from the Clinics, admitted by the Inspectorate, was being overlooked.[152]

Politically active leaders of the Clinics wished to distinguish between the acceptable and unacceptable treatment of addiction to maintain their recently acquired authority. They were hampered by the low status of addiction medicine among hospital doctors, the relatively low level of technical skills required in their field and the limited research on which knowledge could be based and measured against. The GMC's co-opting of criminal concerns as part of its body of ethics in these cases perhaps reflected the need of Clinic elements within and around the Council for an alternative measure of competence as part of their professionalising strategy.

The Dally and Tarnesby cases arose as the focal points of a range of historical forces. Drug use, particularly of opiates, was rising dramatically in the 1980s, along with the number of doctors treating it, particularly outside the Clinics. In response to outside pressure, the GMC had begun to take greater notice of cases concerning doctors' clinical conduct, and more such cases were brought forward for disciplinary action during the 1980s. Its preference for avoiding such matters and for each case to be judged by its members as it arose, was expressed in its distaste for giving specific guidance on appropriate prescribing. Together, these conditions offered a window of opportunity for those doctors who wished to cut through the fog of controversy and exert their authority as the arbiters of proper drug treatment. In the early 21st century, even greater pressures upon the Council and the profession were similarly employed to devastating effect on private prescribers.

5
'Friendly' Visits and 'Evil Men': The Home Office Drugs Inspectorate

There was a sort of general feeling amongst doctors that if you were interviewed by one inspector, you were OK but if two of us turned up you were in trouble. And I think that a number of doctors who were then treating addicts [in the mid-1980s] were sort of advised if they carried on like this they could end up in tribunal and a number of them bailed out from treating addicts, a lot of private doctors said 'I don't need this aggro'.[1]

Introduction

From its origins in 1916[2] to its demise in 2007 the Home Office Drugs Inspectorate gained and then lost great influence over British drugs policy and the regulation of prescribing. The Inspectorate was just one strand in a web of control systems, both state and professional, which emerged over the 20th century to regulate the fate of pharmaceutical 'dangerous drugs', later known as 'controlled drugs'.[3] In addition to the General Medical Council discussed in the previous chapter, this included the police's Chemist Inspecting Officers (CIOs), the Regional Medical Service employed by Health Authorities, the Medicines Control Agency and the Royal Pharmaceutical Society's inspectorate, with professional and state systems working independently and co-operating informally with each other.

The Home Office itself, through a combination of poor archiving practice and refusing the author access to Tribunal documents, provided only a little in documentary material for study, most of it relating to the 1980s. Although reticent to provide documents, the Home Office was generous in granting extensive interviews. Five inspectors, including the last two Chiefs, were interviewed as well as a police Chemist

Inspecting Officer and a former Chief of the Royal Pharmaceutical Society. Bing Spear wrote a historical account of its origins but the rest of his book on the 'British System' took a broader policy scope, leaving the Inspectorate's development somewhat on the sidelines.[4,5] As a result of the source materials available, this chapter gives greatest weight to the 1980s.

The Inspectorate's main concern through most of the 20th century was to prevent diversion of an increasing range of controlled substances from authorised medical channels to unauthorised suppliers or users. It was also responsible for policing the import, export, distribution and manufacture of controlled pharmaceutical drugs.[6] It did this through a staff of inspectors originally based in London and then with additional regional offices.[7] Spear attributed its origins to 'the belief that from time to time it might be necessary to make special enquiries, probably involving medical practitioners, for which it would be better not to employ the police'.[8]

The Inspectorate

Prior to 1868, opium and other psychoactive substances were available for purchase through grocers' shops without any professional or state controls.[9] From the 1868 Pharmacy Act onwards opiates became subject to light but rising professional and state controls.[10] This continued into the early 20th century, with additional substances being brought under control, while opiate use was actually diminishing.[11] America's international influence in the pre-First World War Hague Conventions had led to Britain's reluctant involvement in the development of an international control system for narcotics. These restricted opiate and cocaine to what was described as 'medical and legitimate' use.[12]

The First World War invigorated the British state's interest in narcotics control, as it did in other areas of personal behaviour, including alcohol consumption. Concern about international smuggling and use of cocaine by soldiers culminated in the government designating drugs as a police matter with central controlling authority at the Home Office.[13] Regulation 40B of the Defence of the Realm Act 1916 which resulted from these developments was much stricter than anticipated pre-war. From this point on only medical doctors, pharmacists and veterinary surgeons could possess, give or sell cocaine and opium (although not morphine). It was the Home Office rather than the pharmacy profession who became 'the initiator and arbiter of restriction'.[14] Berridge has characterised these legislative changes as resulting from a combination of press agitation through sensationalised, often inaccurate portrayals of the drug scene, a lack of opposition from leaders of the medical profession, under-representation of grass-roots medical opposition to greater regulation and Home Office leadership favouring a penal approach to drugs.[15,16]

The Inspectorate grew out of the task of monitoring compliance with the Defence of the Realm Act. The Home Secretary gained the power to withdraw a doctor's authority to prescribe cocaine and opiates and in 1917 senior police officers joined Home Office officials in policing these new controls. Pharmacists were now required to keep records for inspection of the prescriptions they dispensed.[17–19] Police officers, who later became designated 'Chemist Inspecting Officers', were responsible for inspecting records maintained by retail pharmacies from 1921 when the Dangerous Drugs Act came into force.[20–22] Underlying these moves against free access to drugs were cultural changes in the early 20th century that, to some extent, diminished the acceptability of opiates and cocaine in society.[23] This can be compared with patterns seen in later settings, such as the USA in the 1980s, where restrictive legislation followed an existing decline in drug use.[24]

By the mid-1920s it was not Home Office Inspectors or the police who inspected the supply of dangerous drugs but medical officials. Regional Medical Officers (RMOs) had been given these powers in 1922, they were employed by the Ministry of Health but available to the Home Office to maintain doctors' compliance with the 1920 Dangerous Drugs Act and Regulations.[25] Doctors gained preferential self-regulation over pharmacists, as according to the Rolleston Report, 'The records kept by wholesale chemists and by pharmacists are inspected by Home Office Inspectors or by the police; but it was considered preferable that those kept by medical practitioners should be inspected by medical officials.'[26] Furthermore, doctors employed by the state as RMOs were not expected to undertake any enquiry that could impair their relationship with GPs such as giving evidence in court against a fellow member of the medical profession.[27]

It is not clear when Home Office inspectors took over the inspection of doctors entirely, but by 1952 RMOs were seldom involved in drug enquiries, with the Inspectorate writing to doctors when pharmacy records revealed they had been prescribing dangerous drugs. The RMOs resumed this work from 1964 until their Service was abolished in 1991.[28,29] Spear attributed these switches from using lay state employees to professional state employees and back to varying workloads of the different parties at given times and to sensitivities around lay and professional expertise in regulation.[30] Certainly these sensitivities arose throughout the 20th century and beyond.

While Spear's name permeates the history of the Inspectorate in the second half of the 20th century, other chief inspectors also left significant imprints on its policies. Charles Jeffrey, Chief Inspector from 1970 to 1977, left few accessible documentary sources for the historian to assess his contribution but seems to have taken a personal approach to the welfare of addicts that has often been credited to Spear alone. Ken Leech, community theologian at St Botolph's, Aldgate, and active with drug users since the 1960s, mentioned that drug users often invited themselves to tea at the Home Office under Jeffrey's leadership and an inspector of the Royal Pharmaceutical Society remembered Jeffrey as a very sociable man.[31,32] For many, though, his memory seems to have been overshadowed by the more charismatic figure of Spear.[33,34]

During Spear's 34 years with the Inspectorate he became the most celebrated civil servant in the drugs field during the 20th century. His posthumously published book, *Heroin Addiction Care and Control*, was prefaced with warm appreciations and he was remarkable in his ability to gain the trust and respect of fiercely divided parties in the treatment and control arenas.[35] His personal concern for addicts and encouragement of doctors to take on their care, including those in the private sector, showed an interest beyond the mechanics of regulating the drug supply. In his last couple of years at the Inspectorate, Donald McIntosh's stricter and less permissive attitude towards private doctors, including some who had previously been visited by Inspectors on a 'friendly' basis, dominated.[36] Dally considered that it was Spear's frequent absences in hospital which 'gave an opportunity to harder and more traditional bureaucrats',[37] but others saw McIntosh as Spear's preferred successor.[38]

Spear complained that the Clinic psychiatrists had 'succeeded in imposing their own ethical and judgemental values on treatment policy',[39] but he himself was far from morally or politically neutral. He was able to sway government policy to modify the influence of the Clinics and he attempted to diversify prescribing and treatment provision for drug users while retaining, and in some cases strengthening, the Home Office's own regulatory mechanisms.

Peter Spurgeon followed Spear as Chief Inspector in 1986, moving straight into the post from criminal policy work within the Home Office. Keenly aware of the respect in which Spear was held both within and outside the Inspectorate, Spurgeon's annual reports suggested a similar approach to his predecessor.[40–42] It is not certain why he, rather than

McIntosh, got the chief post but Spurgeon attributed his appointment from a managerial post outside the Inspectorate, rather than the promotion of an internal candidate, to the tendency towards a more managerial approach across government in the mid-1980s.[43] Certainly his appointment does not seem to have been an attempt to alter the political direction of the Inspectorate. Spear was sympathetic towards drug users, highly critical of the enforcement-dominated US approach, and wary of claims for what could be achieved through policy as 'sooner or later society will have to reach an accommodation with drug use'.[44] Spurgeon's attitudes largely matched these. He had a historical sense of his place within the traditions of the Inspectorate and was happy to follow Spear's 'compassionate approach to the problem',[45,46] but was promoted to a post outside the Inspectorate in 1989.

Like Spurgeon, Alan Macfarlane was a career civil servant with no previous experience in the Inspectorate but his approach to prescribing differed sharply. Taking over in 1990 and staying until his retirement in 2008, Macfarlane was highly critical of private prescribers, the majority of whom he considered 'wicked',[47] and unsympathetic towards the views of drug users themselves.[48] At the end of the 20th century, drug users were organising into activist groups that were beginning to receive recognition from charities and local government,[49,50] but under Macfarlane the Inspectorate took a hostile view to their participation in the policy process.[51]

The success of Dr Adrian Garfoot's appeal after a lengthy and expensive Tribunal was a serious blow to Macfarlane and the Home Office. Macfarlane had chosen the wrong private prescriber to pursue: Garfoot's heavyweight defence fought a far harder battle than anticipated by the Home Office prosecution who were used to easy, uncontested admissions from the accused.[52] Macfarlane was also on the losing side in the plans to extend the Home Office's licensing scheme as recommended by the Department of Health's Clinical Guidelines Working Group. Under Macfarlane's leadership the Inspectorate also lost two important regulatory tools: the Addicts Index and the Tribunal system, while suffering diminishing influence over policy.

In 1970 three mechanisms of state control existed for dealing with prescribing controlled drugs: the Home Office Drugs Inspectorate, the police's CIOs and the Regional Medical Service (see Table 5.1). From 1989 the Medicines Control Agency also inspected pharmaceuticals for quality but was rarely involved in matters of controlled drug prescription and so is not discussed here.[53] The police and the Inspectorate continued through the last decades of the 20th century, with expansion

Table 5.1 Regulatory bodies in the supply of controlled drugs, 1970–1999.

	Home Office[55-58]	Police[59-61]	Royal Pharmaceutical Society (RPS)[62-63]
Inspectors	Drugs Inspectorate	Chemist Inspecting Officers	Pharmacy Inspectorate
Staff	Civil servants	Police officers	Mostly pharmacists employed by the RPS Pharmacies only inspected by pharmacists
Source of regulatory powers	1916 Defence of the Realm Act amendments 1920 Dangerous Drugs Act and 1921 Regulations 1971 Misuse of Drugs Act	1920 Dangerous Drugs Act and 1921 Regulations 1971 Misuse of Drugs Act	Pharmacy and Poisons Act 1933, and Poisons Act 1972 Medicines Act 1968 and Misuse of Drugs Regulations, 1985
Areas of responsibility	Legitimate pharmaceutical industry Illicit drugs industry Medical profession Veterinary and dentistry professions	Pharmacies	Mostly community pharmacies Some hospital pharmacies (those registered with the RPS) Powers to inspect other retail premises where medicines or poisons were sold No responsibility for wholesalers or manufacturers or premises of doctors
Areas of concern	Diversion from legitimate medical use: criminal supply 'irresponsible prescribing' Criminal manufacture	Diversion from legitimate medical use: pharmacies doctors	Professional conduct of pharmacists (Criminal matters other than medicinal matters were referred to police)

Regulatory action	Visits by inspectors. Tribunals (from 1973 to 1997) Action via: • GMC • the courts Not a prosecuting authority itself	Visits by inspectors Action via: • Home Office Inspectorate • GMC • the courts	Visits by inspectors Referral to RPS Disciplinary Committee which could give: • advice • warnings • disciplinary action and removal from RPS register
Inspection of	Prescriptions Pharmacies' controlled drugs registers Doctors Licensed manufacturers/distributors	Pharmacies' controlled drugs register Prescriptions Doctors	Pharmacies and other retailers of pharmaceuticals
Sources of information	Reports from: • doctors • drug users • public • chemist inspecting officers • other inspectorates • Addicts Index (up to 1997)	Own Index Home Office inspectors Other inspectorates	Any concerned professional, the public, police Liaison with chemist inspecting officers and occasionally Home Office Drugs Inspectorate

of the Inspectorate and varying provision of CIOs. Liaison between these agencies was informal.

In the early 1970s, Home Office inspectors visited doctors who were thought to be over-prescribing, sometimes seeking advice from RMOs. From 1970 to 1973, if not dealing with the cases more informally themselves, the Regional Medical Service, in conjunction with the Home Office Inspectors and the police, referred cases of irresponsible prescribing to the General Medical Council (GMC), which was resistant to disciplining such doctors (see previous chapter).[54]

Spear described the Inspectorate's frustration when they lacked a Tribunal system, but also his powers of persuasion in dealing with Dr Brennan, an elderly Portsmouth doctor who had been supplying local heroin addicts with *Diconal*. He 'would have been an ideal candidate for the Tribunal procedures for dealing with "irresponsible prescribing" included in the 1971 Act. But as these did not come into operation until July 1973 there was little the Drugs Inspectorate could do except try to persuade him to be more circumspect in his prescribing. After I "had a word" with Brennan he decided to have nothing further to do with addicts'.[64]

After the Tribunal system was reintroduced, the role of the RMOs diminished but they could continue to advise the Drugs Inspectorate.[65] The focus of RMO enquiries, however, was to establish why the patient needed these drugs – whether for pain relief, which necessitated no further enquiry, or if the patient was addicted, which resulted in monitoring the case.[66,67] This contrasted with the Inspectorate's later interests in the potential of drugs to be resold on the black market, the safety of quantities or formulations to the user, and, in private practice, the ability of the patient to pay doctors' fees.

After many years housed entirely in Central London, the Home Office gained two regional offices in 1974, dividing Britain into the Northern Region, policed from Bradford, covering the North of England and Scotland, the Midland Region, including Wales, the Midlands and the South West of England, with its office in Bristol, and the South East Region based in London. The purpose of regionalising the Inspectorate is not known. Was it intended to respond more efficiently to prescribing outside London or was it perhaps unrelated to drugs issues, for instance a civil service management decision to create jobs outside London? The regionalisation did result in the establishment of meetings for groups of consultants working in the regional Drug Dependency Units, perhaps counterbalancing the dominance of the London Clinics and their expertise.

The 1970s also saw the Inspectorate offer the Home Office as the new venue for the London Consultants Group meetings. The London Consultants Group was composed of (usually consultant) psychiatrists representing the London Clinics and surrounding area and had been meeting since 1968. They had moved from their initial meeting place of the Department of Health due to perceived interference from civil servants in their decision-making. These meetings were attended regularly by either Spear or one of his inspectors who received information on problem prescribers working outside the Clinics and provided advice and information.[68] Inspectors could also advise the London Consultants Group of any difficulties with its own members to be dealt with internally, although this was rare. Unlike doctors outside the Clinics, the Clinic psychiatrists seemed to enjoy the privilege of informal self-regulation while trying to set the standards by which other doctors were judged (see Chapter 6).[69]

The Association of Independent Doctors in Addiction might have performed a similar function. It was set up with Bing Spear's encouragement and initially met with Spear in attendance at the Home Office, until forced to move from government offices. The Association broke up after Ann Dally's second GMC hearing in 1988. Unlike the Clinic doctors, who worked in the same medical hierarchy, private doctors had no interdependency or perceived shared self-interest; the reasons for this are discussed in detail in the next chapter and they never gained the power enjoyed by the London Consultants Group.

As well as encouraging self-regulation among doctors outside the Clinics, the Inspectorate continued to regulate on behalf of the state. In 1985, 228 practitioners were visited by inspectors,[70] increasing from that year. According to one inspector, more visits were made after 1985 because 'it was one of our operational priorities and we were trying to encourage doctors to prescribe responsibly'.[71] This coincided with the appointment of Donald McIntosh as Senior Inspector in the South East Region who raised the number of interviews with private prescribers. Peter Spurgeon, Chief Inspector from 1986 to 1989, claimed that 'in any one year my Inspectors interview some 300 doctors about the safeguards necessary to minimise the risk of diversion'.[72]

The difference between Donald McIntosh's approach towards private prescribers compared with Bing Spear's has been attributed to the disparity between London and Bradford and McIntosh's desire to see equal standards applied in the South East. He was dubious about the role of private prescribers in the treatment field and in his new post launched a campaign to regulate the private prescribers more rigorously.[73] This included writing a report recommending that Dr Dally be taken to a Tribunal. However, Spurgeon saw McIntosh as simply reflecting the established treatment orthodoxy against which Spear, with the exceptional respect he was accorded, was able to speak out.[74]

The Inspectorate was the first arm of British government to develop extensive expertise in drug misuse. Prior to the expansion of research on drug misuse outside government in the 1980s, it was one of the few agencies able to gather data on drug misuse 'on the streets', employing roving inspectors in the years before the proliferation of street agencies for drug users. For this reason it played an important role in providing policy advice to ministers, and constituted a major influence in the regulatory battles between private and NHS prescribers during the period up until the departure of Spear in 1986, after which the Inspectorate's influence waned. The uniqueness of the Drugs Inspectorate should not be overstated though: the role of policy advisor to ministers was one which was also developed by other central government inspectorates, such as the railway inspectors advising the Department of Health on wider transport policy,[75] and accorded with Weber's description of self-perpetuating bureaucracies.[76]

Within the policy community, the Inspectorate gathered and relayed information about all aspects of prescribing controlled drugs. Although perceived, at least during Spear's time, as a neutral force, trusted by all sides, the Inspectorate had its own policy goals. Even at the height of its powers under Spear, the Inspectorate often needed alliances with other interests to push through its desired policies. While at the Inspectorate Spear was circumspect in expressing his views. In his retirement he was more forthright, describing the way the Clinics were implemented as 'an unmitigated disaster'.[77,78] The power of the Clinics' leaders was perhaps reflected in the extent to which the Inspectorate was bound to accept them as setting the standards of acceptable treatment, despite Spear's own views, and Tribunals were never used against Clinic doctors for irresponsible prescribing.[79,80] On the converse side, they sometimes failed to gain policy changes opposed by Spear.

Alongside the Tribunals, the Inspectorate wielded another regulatory tool: the licensing system which controlled who could prescribe certain drugs in the first place, rather than stopping them, as the Tribunal system did, after the fact. Although the Home Office seemed to have the power to decide who received licences, their ability to rescind them was successfully challenged. The Home Office almost exclusively granted the licences to psychiatrists working in the new NHS Clinics and only two or three doctors were ever licensed to prescribe heroin privately.[81,82]

One of them, Dr Kanagaratnam Sathananthan, who was also the consultant psychiatrist at Croydon Drug Dependency Unit, received his licence in the 1980s, probably with Spear's support,[83] and although the Inspectorate later tried to withdraw his licence, the doctor's appeal to the Home Secretary succeeded and he continued to prescribe heroin privately until his retirement.[84,85]

From 1968, there had been a series of unsuccessful attempts originating with the London Clinic consultants to extend the Home Office's licensing powers and further restrict the prescribing powers of doctors outside the Clinic system. The Home Office at first opposed and then supported these moves. It is likely that Spear opposed the extension of licensing in 1984 and certainly the advice given by his department to ministers was intended to dissuade them.[86] A further effort to extend licensing to cover all injectable opiates and restrict licences to 'doctors working in, or under the direct supervision of, a consultant or equivalent in a clinic' was made in 1985 through the Social Services Committee.[87] In its response the government cited misleadingly optimistic trends in prescribing from figures prepared by Spear's department (see Chapter 3).[88-90]

As a key source of advice to ministers on drug policy, Spear's Inspectorate used its position to encourage the policy embodied by the earlier British System and to prevent the Clinics gaining a stranglehold on prescribing. In the 1990s the Inspectorate took a very different view. Under Alan Macfarlane's leadership the Inspectorate made alliances with London consultant psychiatrists and the Department of Health in order to extend opioid licensing and increase its regulatory powers over private prescribers and GPs. Unlike Spear, who was well known for his personal interest in the welfare of drug users and doctors' clinical autonomy, Macfarlane's interest was more heavily weighted towards controlling the drugs supply and preventing diversion, and less to the provision of treatment.

The Misuse of Drugs Tribunal he pursued against private prescriber Adrian Garfoot turned into a fiasco as it wore on from 1993 to 1997 and presumably frustrated by its length and expense, Macfarlane described these procedures as 'cumbersome in the extreme'.[91] Mr Macfarlane saw the preparation of the third edition of the clinical guidelines (1996–99), backed up by another attempt to extend licensing, as an opportunity to streamline tribunal procedures and acquire an enforceable standard for prescribing.[92] Dr Anthony Thorley, the Senior Medical Officer responsible for drugs at the Department of Health in 1996, agreed with Macfarlane on this issue, along with Professor John Strang, Chairman of the Clinical Guidelines Working Group (see Chapter 7).[93,94] Despite some opposition within the Clinical Guidelines Working Group,[95] the principles of the licensing extension were proposed in 1999 by the Working Group in its confidential report to ministers, as intended by the Home Office, Department of Health and Professor Strang. The operational details were then drawn up by the Home Office Drugs Inspectorate and sent to a range of organisations for consultation.[96] This alliance brought the extended licensing proposals further along the path to implementation than ever before but the consultation came to nought and the proposals were never implemented. The Home Office then passed responsibility for the existing licensing system to the Department of Health.

Self-regulation, albeit under the threat of state regulation, seems to have won out. The perceived need for licensing may have been lessened by the Royal College of General Practitioners establishing a 'Certificate in the Management of Drug Misuse' in 2000 to improve levels of training among their members.[97] In 1984 the same licensing proposals may have been dropped due to two factors which were also present in 2001, opposition from GPs keen to guard their clinical autonomy and fears in government that greater restrictions on prescribing would deter reluctant GPs from treating drug users.[98,99]

The history of the licensing issue may also point to the Inspectorate's shrinking influence within the policy community after Spear's departure. Aside from the loss of Spear's deep personal knowledge from the Inspectorate, expertise on drugs had proliferated both outside and inside government independent of the Inspectorate. Furthermore, the cross-departmental alliance pushing for licensing reform may have been stronger than ever before but GPs provided a larger proportion of drug treatment compared with the specialist Clinics and their status as a whole had risen over the intervening years. GPs' prescribing freedoms were not to be withdrawn lightly and government alliances with psychiatry were less effective than they might have been in the late 1960s and 1970s.

Like the General Medical Council discussed in the previous chapter, the Home Office had difficulties in defining what kind of prescribing was medically legitimate: Did this include drugs purely to satisfy the cravings of addiction? Did it cover sending prescriptions in the post to patients not seen for long periods of time? In the early years of the Inspectorate, there had been uncertainty over whether a patient who had originally been prescribed drugs for a medical condition and had become dependent upon them once the medical condition had passed should still receive them merely for the relief of addiction.

According to the 1926 Rolleston Report, which endorsed a medical approach to addiction, this uncertainty over matters which 'must turn largely on questions of medical opinion' made the Home Office reluctant either to prosecute doctors or to bring a case to the GMC for conduct 'infamous in a professional respect'.[100] The all-medical Rolleston Committee advised that for these cases, special medical Tribunals be set up so that a doctor could be judged by his peers instead of the courts.[101] These Tribunals again maintained the idea of exclusive professional expertise considered by the Rolleston Committee to be lacking in a lay magistrate.[102] However, the provisions were never used and then removed in 1953 pending agreement with the medical profession about new procedures.[103]

Avoiding the courts had obvious advantages for a doctor, who could not be given a criminal record, fined or imprisoned by a Tribunal. Furthermore, the Home Office Tribunals would be held in private, while the GMC hearings were open to the press and public. Before 1970, when the penalty of suspension of registration was introduced,[104] the GMC's only sanction was the drastic one of erasure from the register. By contrast the harshest penalty the Home Office could apply was removal of the right to prescribe dangerous drugs, leaving the doctor still able to practice most areas of medicine.

The Tribunal system was not reintroduced until 1973 under the Misuse of Drugs Act, 1971.[105] Under this Act, directives by the Home Secretary could be applied for criminal offences, most of which concerned dependent doctors diverting supplies for their own use, and for non-criminal prescribing issues. Spear saw Tribunals as essential, 'plugging this gaping hole in our control machinery'[106] and credited it to the Amphetamines Sub-committee of the Advisory Committee on Drug Dependence which had recommended bringing back the system. Tribunals were considered necessary by parliament because the government accepted the GMC's complaint that its own machinery was inadequate and therefore declined to discipline irresponsible prescribers itself.[107-109] As shown in the previous chapter, the GMC was later to become more enthusiastic in prosecuting such prescribers, despite the lack of a relevant change in its jurisdiction or the addition of any surveillance function.

Injudicious or irresponsible prescribing was defined during the passage of the Misuse of Drugs Act as 'careless or negligent prescribing or unduly liberal prescribing with bona fide intent'.[110] The law itself contained no definitions, and the GMC's similarly vague statement 'the prescription or supply of drugs of dependence otherwise than in the course of bona fide treatment'[111] enabled both regulatory authorities to interpret the term subjectively and according to the changing trends in treatment.

The Tribunal panel itself consisted of four medical members nominated by medical bodies: the Royal Colleges, the GMC or the British Medical Association and a Queen's Council barrister acting as chairman. The format of these hidden proceedings was similar to a law court: a lawyer each representing the doctor and the Home Office, with evidence being presented and cross-examined. Tribunals could be used in a variety of ways: the threat of proceedings could persuade doctors to change their practices, as suggested in the quotation opening the chapter and also with Dr Dally in 1986;[112] doctors summoned to a Tribunal might remove themselves from the medical register before it got underway to avoid the ordeal itself; or the doctor undergoing the Tribunal might be acquitted or have their prescribing powers modified by the Secretary of State.[113] A successful prosecution could also be appealed. Between 1973 and 1999 the system was used only once against a Clinic doctor and this was for criminal offences relating to the supply of drugs rather than irresponsible prescribing.[114,115]

It was the case of Dr Adrian Garfoot which brought the whole system to an end. A private GP who had worked with drug users for several years, Dr Garfoot had developed views similar to Dr Dally's regarding reform of the drug control system and what he saw as the oppressed position of drug users in society. Although the Home Office, after several years' delay, managed to prove its charges during the Tribunal, a successful appeal on procedural grounds by Garfoot's lawyers overturned the ruling in 1997. After this humiliating and costly defeat Tribunals were never used again. The Home Office explained its official reasons:

It became clear during the 1990s that these powers [under sections 13–16 of the Misuse of Drugs Act] were no longer an effective mechanism and the last case was referred for Tribunal action in 1993. The practitioner involved, Dr [Garfoot], was able to delay the hearing for over a year, by which time he had engaged other doctors to undertake prescribing at his clinic. Subsequently it has become apparent that the legislation is deficient in several aspects of Human Rights, thereby removing any remaining possibility that the powers could be reactivated.[116]

This returned the Inspectorate to its predicament in the early 1970s, able to advise doctors and gather evidence but without its own disciplinary function. Yet this time the GMC took a much more active approach to regulating prescribers, taking non-Clinic doctors to its Professional Conduct Committee and by the turn of the 20th century was erasing quite a number of private prescribers from its register, including Dr Garfoot in 2001. With a reinvigorated GMC, the Inspectorate reduced its work with doctors, cutting down the number of interviews to 'a handful'.[117] Although there remained co-operation between the state and the profession in regulating non-Clinic doctors, the processes were more weighted towards self-regulation, albeit under the watchful eye of politicians.

A third tool used by the Inspectorate was the Addicts Index, a list of patients believed dependent on opiates or cocaine who were known to the Home Office. The Drugs Inspectorate was the 'custodian and principal user' of the Addicts Index, which had been kept centrally as a formal record since 1934, probably at the request of the Opium Advisory Committee of the League of Nations in 1930.[118] Inspectors also had a role in ensuring that doctors were notifying patients dependent on opiates or cocaine to the Index, as they were legally obliged to do from 1968. The name, drug(s) of addiction and any controlled drugs prescribed were listed on the Index so that, in theory, any doctor prescribing to the notified addict could check with the Index to see whether they were already receiving a prescribed supply from another source, and so prevent patients from 'double scripting'. In practice there was often a long delay between notification by a doctor and entry of the data onto the Index (about three months in 1982), with computerisation in the 1980s only adding to these difficulties.[119]

The Index was closed in 1997 as a cost-cutting measure, against the advice of the Advisory Council on the Misuse of Drugs (ACMD),[120] and statistical information on drug users in treatment was gathered instead from non-compulsory notifications to the Regional Drug Misuse Databases without the names of patients and published centrally by the Department of Health. The end of the Addicts Index meant the loss not only of a source of statistical data on trends in drug use and treatment but also a window onto doctors' prescribing and an early warning system for any new doctor who might require a visit.

Until the 1990s the Inspectorate had been a 'one stop shop' for drug policy, legislation and enforcement[121] but as other parts of government grew, such as the Department of Health, they took over many of its functions and some were cut to reduce costs. By the end of the 20th century the Inspectorate had lost its Tribunal system, the Addicts Index and soon it was to lose its licensing system as well. It was impossible for the Inspectorate to visit the many GPs involved in treatment and without its tools it could do little to regulate doctors. Its remains were wound down between 2004 and 2007, becoming the Drug Licensing and Compliance Unit on 1st January 2009.

While the Inspectorate and police provided state supervision over prescribed controlled drugs, the Royal Pharmaceutical Society and the GMC regulated their own professional members. In relation to the bodies discussed here, the Council's main role was to prosecute, through its disciplinary procedures, cases brought to light by the Home Office Inspectorate, the CIOs, or occasionally, the Royal Pharmaceutical Society's inspectors. It also suggested 'experts' from the Clinics to advise CIOs about acceptable prescribing, at least from the mid-1980s onwards.[122]

Between 1968 and 2010, there were fluctuations in the levels and leadership of disciplinary action against doctors taken by both the GMC and the Inspectorate. Until 1968, the GMC had dealt with a mere handful of cases of 'non-bona fide prescribing' of controlled drugs. After that date, the numbers increased, but remained at less than ten per year until at least 1990.[123] From 1968 to 1973, the GMC was the only regulatory body able to take disciplinary action for 'irresponsible prescribing', although warnings could be given by the Inspectorate and the RMOs (see above). It seems that during the 1980s and for much of the 1990s, the Inspectorate took the lead in disciplinary cases against doctors prescribing controlled drugs.[124,125] However, after the 1997 'watershed' of Adrian Garfoot's successful appeal against his 1994 tribunal ruling, the GMC took over the job of prosecuting all the cases made by the Inspectorate's investigations and the Tribunal machinery was left unused. In the 21st century, the Council became the sole disciplinary body for prescribing doctors.

Internal and external expertise

The Inspectorate's work focused entirely on controlled drugs rather than the whole range of doctors' professional behaviour and it developed its own internal expertise and views on appropriate prescribing and the implications for the demand and supply of both pharmaceutical and trafficked drugs. These included the particular formulations and substances likely to be diverted, their black market values, and the health risks particular drugs posed when not used as prescribed. However, the Inspectorate never employed any doctors or pharmacists and as lay inspectors without medical or pharmaceutical training, they were keenly aware of the sensitivity of commenting on the well-defended turf of doctors' clinical judgement, relying upon external sources of medical advice to support and legitimise their judgements.

While there was no official definition of the 'irresponsible' prescribing that the Inspectors were supposed to police, they had drawn up their own guidelines on what to look for when visiting prescribers.[126] During the 1980s, advice was sought from some of the London consultants and their publications.[127] In 1984, South East Regional inspector John Lawson, was asked at a GMC hearing what features of a particular doctor's prescribing led to the setting up of a Misuse of Drugs Tribunal. He replied that it was the amount of drugs prescribed for individual patients but that they did 'seek expert advice' on that.[128] After 1984, the Inspectorate could also use the medically authored Department of Health Clinical Guidelines.[129–132]

However, when considering whether to take Dr Dally before a Tribunal in 1983, the Inspectorate decided against, while the GMC pursued what was a rather weak case amid attacks in the medical press on private prescribing by London Clinic psychiatrists (see Chapter 4).[133] The GMC, in its self-regulation model, was free to make clinical judgements, relying on its own members for guidance, some of whom were particular critics of private prescribing.

Describing her experience of visits from Chief and Senior Inspectors Bing Spear and John Lawson, Ann Dally recalled, 'I learned far more from them than from so-called specialists or from the medical literature. I tried not to say this to them because it embarrassed them. They were not supposed to be regarded as "medical experts".'[134] Spear in particular was one of the most knowledgeable individuals about the 'drug scene' and prescribing during his time at the Home Office.[135–137]

Advising doctors involved an intricate dance for the Inspectorate, unable to tell clinicians how to treat their patients, using professional peer opinion to justify their advice and yet holding over prescribers the threat of judging their behaviour. Here, Don McIntosh, on an inspector's visit Dr Dally in December 1985, advised her that a former patient from the North of England should, if he needed further treatment, receive it locally and not from Dr Dally. This was the exchange noted,

AD: *I would be reluctant to take him back if you disapprove.*
DM: *It is not for us to approve or disapprove. You could be criticised by your medical colleagues . . .*[138]

Inspectors had to give the appearance of not dictating acceptable treatments to a doctor, yet it was the Inspectorate who would refer cases to a Tribunal. The Tribunal itself was medical in membership but the evidence would be supplied by the Inspectorate. The medical profession therefore retained an overarching power over the Inspectorate's regulatory authority and each was dependent on the other to achieve a disciplinary result.

The publication of the first clinical guidelines by the Department of Health's Medical Working Group in 1984 assisted the Inspectorate by providing further official medical support but did not fundamentally change its approach. The Inspectorate had its own internal view on prescribing before the good practice guidelines appeared. Rather than accepting the advice of the guidelines wholesale, the document was used by the Inspectorate to increase its leverage in enforcing its existing views when visiting prescribers.[139] One inspector commented, 'Once we had some guidelines we could actually point to something, say "You should do this, you should do that, your colleagues said all that."'[140]

The Inspectorate's own internal guidelines, which were drawn up in the early 1980s,[141] dealt mostly with pragmatic procedural matters, but included an appendix which showed what inspectors were looking for. The thrust of the questions, such as 'What steps did the doctor take to satisfy himself that the patient was addicted?' aimed at finding out whether a doctor was willing to prescribe drugs regardless of the patient's condition, and so potentially act as a supplier of drugs for non-medical reasons and not in the treatment of addiction. A checklist of practices indicative of appropriate or inappropriate prescribing included whether the doctor had conducted a complete physical examination, taken blood and urine tests, observed withdrawal symptoms and self-administration of drugs and made various checks on identity and the patient's claims before prescribing.[142]

These criteria show that the Inspectorate's guidelines were quite different in intention from the 1984 good practice guidelines. As it claimed, the purpose of investigations, was 'not to stop a doctor from prescribing controlled drugs to addicts if that is being done in a controlled and responsible manner, nor to force him to conform to a particular treatment regime although advice about consensus trends in treatment may be offered in conjunction with the names and locations of specialist treatment facilities'.[143] This mention of 'consensus' was rather surprising, as it was clearly lacking in the medical profession at that time. When discussing Tribunals in 1986, the Home Office was clearer about different approaches. After taking over from Bing Spear as Chief Inspector, Peter Spurgeon wrote:

> Neither the pursuit of individual cases, nor the Home Office policy underlying the Inspectorate's general approach is coloured by one medical school of thought or another on prescribing philosophy, which remains a highly variable commodity ranging from strict non-prescribing of substitutes in some areas through to open acceptance of long-term maintenance prescription as a last resort in others. Our basic concern is quite simply to ensure as far as is practicable and reasonable that the styles adopted by medical practitioners in their treatment of drug misusers are consistent with the need to prevent significant leakage of controlled drugs into the illicit market.[144]

The extent to which the Inspectorate expressed views independent of the Clinic treatment orthodoxy varied over the period. The early to mid-1970s was a time of experimentation in the Clinics and their leadership expressed little concern about private prescribing.[145] As lines of allegiance hardened, Spear stood out against the methadone and short-term detoxification model held up by the Clinic leaders of the late 1970s and 1980s. According to his successor, Peter Spurgeon, Spear was only able to do this because of the special respect he had built up within the policy community, and the length of his service, joining the Inspectorate in 1952 long before the Clinics existed.[146]

When Donald McIntosh acted in Spear's place in 1985–86, he took a key role in advancing regulatory action against Ann Dally and at her GMC hearing reported asking her, 'Does not long-term prescribing give a soft option to carry on taking drugs?',[147] and suggesting that the fact that her patients were unwilling to go to their local NHS Clinics indicated that they were only coming to her for a supply of drugs. His questions seemed to reveal a view that prescribing was 'perpetuating' patients' 'addiction and problems'.[148]

By the 1990s, the Inspectorate's enforcement policies took into account the longer term prescribing patterns that had emerged from the harm-reduction movement following HIV/AIDS.[149] However, in some respects they could be said to have reflected the aims of the London Clinics, many of which, although influenced by harm reduction, remained resistant to the idea of long-term prescribing.[150] Despite the apparent widespread acceptance of oral methadone maintenance during the 1990s, the overall aim of reducing patients' prescribed doses continued to appear in criticisms of doctors outside the Clinics. Dr Adrian Garfoot's Tribunal in 1994 charged him with prescribing 'without instituting a reducing regime' and not according to the (1991) clinical guidelines. A further charge was that he had not consulted 'experts in the treatment of drug misuse, such as a local drug dependency unit'.[151]

The motives and effects of regulation

Although the Inspectorate's mission was the prevention of drugs being diverted from the authorised channels throughout this period,[152] it pursued additional policy goals and priorities, being guided by its changing internal views on the needs of patients, appropriate treatment and the doctors who provided it. Concern about the health of patients, as well as the destination of prescribed drugs, influenced regulation. When Inspector John Lawson visited Dr Tarnesby's practice in 1981, he claimed to have advised him that prescribing *Diconal* to drug users could be dangerous as this oral tablet was often hazardously crushed and injected.[153] A number of sources also revealed that the Home Office placed emphasis on the motivations of drug doctors, rather than simply the type of prescribing they were undertaking, suggesting an interest in the care given to addicts going beyond their remit of controlling diversion and a degree of moral judgement.

Charles Jeffrey, writing in 1970, referred to two types of drug doctors on the scene a decade earlier: 'script doctors' and a new kind of 'dedicated practitioners' whose motives were 'unimpeachable'; although despite their different motivations, Jeffrey attributed to both the overflow of drugs onto the black market.[154] Spear maintained this distinction,[155] and according to an inspector with 20 years' experience, Spear directed the powers of the Inspectorate accordingly:

> I think Dally was there to help people in the same way that Garfoot was there. They were a bit misguided, but that was what they were doing. Whereas [Dr. X] and [Dr. Y] were just evil men … They were different type of people … [Dr. X] and [Dr. Y] were being taken before Tribunal – a bit evil and in it for the money. There were other doctors prescribing to addicts who were seen on a regular basis but more in a friendly advisory way and Dally was one of these. She was encouraged to get involved by Bing.[156,157]

Although a Tribunal was eventually brought against Dr Garfoot, it was after many years of advice and 13 oral and written warnings given between 1982 and 1992.[158] Similarly with Dr Dally, although a Tribunal was threatened in 1986, seven years after she accepted her first drug addict patient, it was never brought. By contrast 'Dr X' above was served with Tribunal papers only a year and a half after his first addiction prescribing. Theoretically this also applied with the GMC where 'bona fide' intention was considered in prescribing but in practice good intentions, even where proven as far as they could be, could be disregarded.

The regular 'friendly' visits to doctors were not just advisory but also to gather intelligence, at least during the 1980s. Spear remarked,

> Not all visits to practitioners … are in respect of some offences or irresponsible prescribing. There are a few general practitioners who are taking a particularly keen interest in drug misuse problems and who welcome periodical visits from the Inspectorate. In turn much valuable information about the local drug scene is obtained from these practitioners.[159]

Of course under a change of leadership a doctor at one time considered in this intelligence capacity, like Ann Dally, could become one of the formally regulated.

Although the Inspectorate directed the practice of doctors towards controlling the drug supply, it could also be seen as having a training role for practitioners new to the field in a dearth of other sources. *Treatment and Rehabilitation* (1982) remarked on the lack of training opportunities for doctors faced with addicted patients.[160] During the 1970s and 1980s very little time was spent on addiction in the undergraduate medical curriculum nor were there many training opportunities for medical postgraduates other than psychiatrists specialising in addiction. Testifying before the Social Services Select Committee in 1985, Dr Stuart Carne, Senior Tutor in General Practice at the Royal Postgraduate Medical School, agreed that the basic GP training was not sufficient for a GP to be able to recognise an addict. Dr John Cohen, a GP member of the first Guidelines Committee and a Senior Lecturer in General Practice at Middlesex Hospital Medical School, considered there were insufficient experienced psychiatrists in drugs to provide a network for training.[161] Some doctors started to treat addiction with minimal knowledge of treatment modalities and the Inspectorate could be the most knowledgeable source available to them.

While the Clinics were largely left by the Inspectorate to self-regulate, the Home Office extended and delegated policing of the drug supply to doctors working outside the Clinics. The GMC case against Dr Tarnesby, at which the Inspectorate gave evidence for the prosecution, showed a range of issues doctors were supposed to be aware of to maintain control over the drugs supply, aside from and sometimes in potential conflict with the doctor's own perceptions of the patients' needs. For instance the patient should be of known provenance, with a referral from a GP or other doctor. Before the case Spear had written to Tarnesby reminding him of

> the need for extreme caution in prescribing for patients previously unknown to the practice who claim, but cannot readily confirm, that they have been in regular receipt of controlled drugs. As you no doubt appreciate from your recent experience, a doctor who is prepared to accept such patients may soon find himself inundated by similar requests and may well unwittingly become an important source of drugs circulating in the illicit market.[162]

Tarnesby, although clearly trying to make a good account of himself under cross-examination before the GMC, described the change in his practice resulting from this regulatory attention:

> ...when there was a GP I did enquire and the difficulty about it is when the patient states he does not have a GP, and I then thought if he has not got a GP and he says he has never attended anybody for treatment what can I do? But nowadays I would say: 'Then in that case I will not accept him.' Indeed, if now I were faced with that same choice I would say: 'Well, it is just too bad. I cannot accept him', but at that time I felt I must bend over backwards to accept him, and that, I think, can lead to undesirable results, and I would not do it again.[163]

Whether this was accurate or not, it indicated the direction of the pressures on these doctors. On the question of patients losing their jobs once in treatment, Mr McIntosh, when giving evidence against Ann Dally, explained 'it was incumbent upon her to make the most stringent continuous inquiries to satisfy herself that this person had a legitimate means of meeting the costs without resort to some criminal activities'. And although not explicit, it seemed expected that a private prescriber should discharge any patient found to have lost their job.[164] Tarnesby claimed that most of his patients owed money but were still seen.[165]

Doctors were also discouraged from accepting patients outside the locality of their practice, as the Inspectorate was concerned about the geographical spread of patients, not as a treatment or medical issue, but concerning the market in diverted drugs. John Lawson, an Inspector giving evidence at Tarnesby's GMC hearing, stated, 'Most doctors who prescribe for addicts tend to attract addicts from their own area. Once a doctor starts to attract addicts from other parts, the whole of London, or the Home Counties, we become suspicious that he is a "soft touch".'[166] The Inspectorate was also concerned that, through 'long-distance prescribing' markets in diverted prescribed drugs could develop in areas outside London that had been previously unaffected.[167]

An unexpected role of the Inspectorate was in finding doctors for patients such as those whose doctor had been disciplined and could no longer prescribe. The Home Office directed Dr Dally's patients to both an NHS Clinic and another private prescriber after her second case.[168,169] According to Spurgeon, this function was a result of 'the relationship with the drug using community built up by Bing Spear', but may have preceded his tenure. The criteria for choosing these doctors for referrals were, according to Spurgeon who could not be seen to show preference for particular treatment modalities, 'logistics and practicalities' rather than treatment modalities.[170] After Spear's departure it is not clear how long this practice continued.

When visiting doctors, the Inspectorate not only expressed its own concerns, but also made them aware of the interests of the press, particularly the tabloids, in their practices. In this, the popular press acted as an additional regulatory pressure upon doctors, particularly those in private practice. The tabloids targeted private prescribers in the 1980s and 1990s as a source of scandal which also yielded information of varying reliability for the Inspectorate and GMC.

Inspector John Lawson reported that in a conversation in December 1981 'I mentioned to Dr Tarnesby in my experience he would have to be careful when dealing with addicts because the press once they became aware a doctor is dealing with addicts can see headlines and they are apt to put in reporters claiming to be addicts.'[171] Despite this warning Dr Tarnesby was hoodwinked by a *Daily Mirror* reporter posing as a drug-dependent patient to whom he prescribed. After this Tarnesby claimed to have instituted greater checks such as physical examinations of the patient to check for signs of injecting, and 'I decided I will never use self-injectables again, and never did',[172] although this latter claim was disputed by another inspector.[173]

Alerted by the reporter's article,[174] the Inspectorate interviewed Dr Tarnesby after the journalists' accusatory article appeared in the *Daily Mirror*, but were satisfied with his answers and did not pursue the matter further.[175] In Tarnesby's GMC hearing it was not made explicit why the Inspectorate had dropped the matter but cross-examination of the Home Office inspector and the reporter suggested that some aspects of the newspaper's account might have been fabricated.[176]

The workings of the Inspectorate can be seen as those of a Weberian bureaucracy whose power was based in technical expertise and knowledge developed through experience in the service. The Drugs Inspectorate developed into a source of policy advice for government, training for prescribers, an occasional referral agency for patients, and policy actor in its own right, trying to maintain itself as a permanent institution rather than serving the ends for which it was originally designed.[177] Rhodes observed similar processes in his wider examination of inspectorates within British government, which often surpassed their original task of inspecting. Rhodes found that other central government inspectorates, as well as enforcing legislation among those they inspected, had developed into professional advisors to ministers and departments.[178] He saw inspectorates as not only enforcing standards, but setting them too, matching the Drugs Inspectorate case. He also found inspectors dividing those they visited into the reputable, inspected on a friendly basis, and the dishonest, who were prosecuted.

The act of surveillance has received particular attention in the history of medicine and beyond since Michel Foucault's ideas on the place of the body in modern medicine and on 'disciplinary power' became influential.[179,180] However, the impersonal nature of Foucault's surveillance did not fit the very individual imprint left by the Inspectorate's changing leadership or the personal relationships between observer and observed. Furthermore Foucault's denial of personal agency as a historical force has been hard to square with this picture. However, Foucault did not necessarily intend his ideas to be taken as a general or consistent theory or to be applied to other historical contexts.[181] His followers were more imperialist in their claims and some of their work may inform this one. David Armstrong's expansion on Foucault's ideas to medical surveillance in the 20th century has provided an interesting comparison. He described the archetype of a tuberculosis 'Dispensary' which acted as a central clearing house for information about sickness and potential sickness in the wider community, mapping the spread of disease and gaining the consent of the well population to undergo policing and surveillance.[182] This contrasted with the institutionalised surveillance of prisons and schools in that it looked into the spaces between bodies in their community environments creating a new concept of social space taking groups of people to be the reservoirs of disease.

The infectious disease model of drug addiction that was overtly expressed to justify the compulsory notification requirements for the Addicts Index, a key tool for the Inspectorate's surveillance of doctors and patients, showed clear parallels to Armstrong's Dispensary. Like the Dispensary, the Inspectorate was a central clearing house for information and intervention among the community, in this case one made up of doctors and drug users, to varying degrees of consent. The movement of drugs, which could equate with agents of infection in the communicable disease model of the second Brain Committee, between patients and other drug users, had taken the place of the tuberculosis bacilli. However, although Armstrong's model of the Dispensary may have illuminated the development of social space in infectious disease, it has not substantially added to understanding the Home Office Inspectorate or the Addicts Index, the roles of which were openly declared to be part of a public health system of control of drug addiction in which healthcare and disciplinary processes were combined.

Conclusion

Throughout the last decades of the 20th century, the Inspectorate's regulatory gaze fell on the doctors working outside the Clinics despite evidence that drugs were also leaking from the Clinics onto the illicit market.[183] While the Clinics' leaders united successfully to largely self-regulate and were becoming integrated into establishment institutions, private prescribers remained fragmented and politically weak. GPs strengthened their status over the period and by the end of the century fended off further state regulation. The early twenty-first century move of much drug treatment into voluntary services, albeit paid for by government, allowed many GPs to step aside from the treatment of addiction.

The Inspectorate's regulatory tools and how they were used passed through several different phases over the period: *1970–73* was a period of frustration; the Tribunal system passed into legislation in 1971 but could not be used for another two years, while the GMC was reluctant to take action itself; from 1973 to 1982 the Tribunal system was used but only occasionally probably due to unwillingness by Home Office lawyers to accuse doctors of irresponsible prescribing while the GMC took some action itself.[184] After 1982, the GMC continued to discipline private prescribers and Tribunals were used more frequently until the mid-1990s Garfoot fiasco and in 1997, the Inspectorate also lost the Addicts Index.

Macfarlane attempted to find a new tool to replace the Tribunals by making the 1999 guidelines enforceable through an extended licensing system but this too failed (see Chapter 7). The Inspectorate's work with doctors shrank and the GMC took over as sole prosecutor of 'irresponsible' prescribers, gathering momentum at the turn of the century. As the Inspectorate became increasingly ineffectual, it was ripe for further cost-cutting by the Home Office who either eradicated its functions or shifted them over to the Department of Health. For a short time the Inspectorate continued to co-operate with the GMC by providing some of the information used for its cases. Despite the demise of the Inspectorate and the increased activity of the GMC, particularly over private prescribers, the combination of state and professional regulation continued. Government and state-sponsored guidelines kept arriving on doctors' desks and the GMC was kept aware of the wishes of government.

Until the mid-1980s the Inspectorate, like some other central government inspectorates, played a key role within the policy community, both in advising ministers and other policy bodies such as the ACMD, and also by supporting and protecting doctors who differed in philosophy from the London Clinics but which it judged to be well-motivated. Under Spear, the Inspectorate worked to maintain diversity in treatment services by opposing the extension of licensing in the mid-1980s. Bing Spear was particularly influential because of his own highly respected knowledge, personal charisma, perceived neutrality and because of the narrower policy community of his time. Spear fostered the internal expertise of the Inspectorate, mostly using medical advice to support and legitimise existing lay-developed policy. His long service with the Inspectorate, which pre-dated the Clinics', may have strengthened his position when opposing their policies, something lacking in his successors.

Once Spear had gone, there appeared to be an opportunity for the Clinics to strengthen their position and curtail the prescribing of other doctors. After manoeuvres against private prescribers initiated by

McIntosh, policy was pushed in the opposite direction by events outside the Inspectorate. The emergence of a near consensus for harm-reduction treatment policies which developed after HIV/AIDS became a policy issue, and the diversification of the policy community to include more non-medical influences weakened claims of the Clinics to be the sole source of expertise. This provided opportunities to those pursuing a more liberal prescribing policy outside the Clinics.

During the 1990s the Inspectorate's leadership sought greater control over non-Clinic prescribers but failed to hold on to two sources of strength, the Addicts Index and the Tribunal system and its powers faded. In the early 21st century the Home Office leadership was in a strong position relative to the Department of Health[185,186] and this may have enabled its leadership to 'cherry-pick' the functions it wanted to keep or pass on to the Department of Health. After shedding the Inspectorate's prescribing licensing to the Department of Health, it also left behind the rest of its work regulating doctors.

Meanwhile, the GMC was being pressed to increase its activity not only by government but also by the media. By contrast, the Inspectorate was noteworthy in its low public profile and was rarely heard of outside the drugs field. This meant that, unlike some other central government inspectorates, public opinion had little influence on its policing priorities. Drug policy in the 1960s and 1970s had been largely determined behind closed doors between civil servants and members of the policy community. From the mid-1980s, public opinion and political interest had a little more influence, although in a scattered and inconsistent fashion but the priorities of the Inspectorate continued to reflect its own internal views and elements of the policy community until the end of the century. Greater public scrutiny was brought to bear on the Inspectorate during the Shipman Inquiry which had been set up to investigate the regulatory failures that allowed GP Harold Shipman to murder 15 of his patients using controlled drugs. The Inspectorate escaped much of the criticism for these failures but in 2004 the Shipman Inquiry recommended the creation of a new multi-disciplinary inspectorate which could take over some of the work of the Home Office Drugs Inspectorate.[187] The government rejected this idea but abolished the existing Inspectorate regardless.[188]

During its lifetime, the Inspectorate informally co-operated with the other strands in the regulatory network, both state and professional, to gather intelligence and to advise and discipline those non-Clinic doctors it found wanting. To influence policy according to its own agenda, the Inspectorate made strategic alliances with medical professionals, other government departments and policy bodies, with varying degrees of success. This was especially important for achieving acceptance as a non-medical body regulating doctors. It was not the cold, calculating eye of Foucauldian surveillance but played a very personalised role in the public-private struggle. The face-to-face relationships between the surveillant and surveilled and the changes in the Inspectorate's leadership greatly influenced its methods and policies.

6
Unifying Hierarchs and Fragmenting Individualists: Three Professional Groups

Introduction

After Ann Dally left the drugs scene, her organisation, the Association of Independent Doctors in Addiction (AIDA), collapsed; but its short life and that of two other doctors' groups can help us answer a question central to *The Politics of Addiction*. Why were the London consultants able to fend off outside regulation and set the standards by which other doctors were judged while the private doctors succumbed to extensive discipline? The Association of Independent Doctors in Addiction (1981–88), along with the Association of Independent Prescribers (AIP) (1996–98) and the London Consultants Group (LCG) (1968 to the present) represented private prescribers, NHS GPs and the London Clinic doctors. All three groups were acting as less formal mechanisms of self-regulation than the Inspectorate or General Medical Council (GMC). By comparing the characteristics and histories of these three groupings, I hope to show how and why the strategies of each succeeded or failed in protecting their own interests.

To facilitate comparison, I use the system of classification known as Cultural Theory. Originally developed by anthropologist Mary Douglas,[1] Cultural Theory has linked values and beliefs to social relationships, and from these, has explained behaviour. Summarised elegantly by David Oldroyd, people are classified first, 'according to their degree of commitment to the social *group* to which they belong and by which their actions are shaped and determined; and second, according to the intensity of social control to which they are subjected by reason of the social categories and concomitant roles that obtain within the society of which they are members i.e. *grid* control'.[2]

117

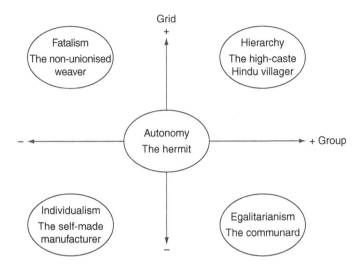

Figure 6.1   Five archetypes mapped onto the two dimensions of social structure. Reproduced with permission from M. Thompson, R. Ellis and A. Wildavsky, *Cultural Theory* (Boulder, CO and Oxford: Westview Press, 1990), p. 8

These dimensions can be represented as two axes, as in Figure 6.1. At the 'zero' position along the 'group' (*x*) axis, the individual was in a network of his own making which had no recognisable boundaries. Others, further along from this position, may have belonged to several associations which were clearly bounded so that they could determine who was and who was not a member. The more an individual's life was absorbed in the group, for instance working inside the group, marrying inside the group and so on, the stronger their 'group' score would be.

If social categories of people and their appropriate behaviour were heavily imposed by a culture, then 'grid' was stronger; if behaviour and status were more flexible or left to individual autonomy, then grid was weaker. In combination, these dimensions have produced five possible social forms: hierarchy (strong grid, strong group), egalitarianism (weak grid, strong group), fatalism (strong grid, weak group), individualism (weak grid, weak group) and autonomy (zero grid, zero group). These archetypes were extremes, perhaps never found in actual existence but useful as explanatory tools. To illustrate these archetypes, Thompson, Ellis and Wildavsky have given the examples of the hierarchical high-caste Hindu villager, the egalitarian communard, the fatalistic non-unionised weaver, the individualistic self-made manufacturer, and the autonomous hermit (see Figure 6.1).

The 'strong grid' high-caste Hindu villager and the non-unionised mill worker were both constrained by a socially imposed 'gridiron' of things they could and could not do, but while the villager was part of a larger hierarchical group which gave him rights to land and deference from those beneath him, the non-unionised mill worker was isolated from other workers and experienced no solidarity with them, lacking also any scope for competition.[3]

The 'weak grid' self-made manufacturer and the self-sufficient communard both considered themselves much freer to act as they pleased, one to hire and fire, and the other to act as equal to his fellow communards, uncontrolled by the perceived coercive world outside the commune. The self-made manufacturer got where he was through rugged individualism, valuing market mechanisms, and using individualistic and pragmatic strategies through networks he had developed himself. The communard was defined through membership of a group that rejected the inequalities of the outside world. The commune's only principle of organisation was rejection of those outside the group's boundary; there were no set ways of resolving conflict or reaching decisions inside the commune.

Last of all was the hermit, who was not necessarily reclusive but withdrew from the coercive social involvement of which the other four types became part. The hermit valued autonomy above all else, and aimed at a life of relaxed, unbeholden self-sufficiency, trying to avoid both the manipulations experienced by the mill worker and the communard, and opportunities for manipulation of others open to the manufacturer and high-caste Hindu. His job might have been driving a taxi, working alone, with ambitions only to be self-sufficient rather than expanding business to work with others.[4]

Corresponding with all these differences, were value systems and strategies relating to all aspects of life, including attitudes to authority, working to long-term and short-term goals and the control of information. People classified in these categories were not conceived of as lifeless automatons, but able to think critically about their situations. The contexts in which they lived were not rigid structures but constantly re-created by individual actions: they were the results of myriad individual decisions made in the past and re-shaped each day.[5]

To investigate the three doctors' groups, I interviewed members of each as well as civil servants from the Department of Health and Home Office and a manager from a drug action team involved in local policy priorities. The main documentary sources for this chapter have been the committee papers and associated correspondence of AIDA and the AIP.

Those of AIDA were deposited by Dr Dally at the Wellcome Library for the History and Understanding of Medicine, while one of the founders of the AIP shared the Association's papers. Documents from the LCG proved much more elusive. They were sought through a number of routes throughout the research project but their existence was repeatedly denied. Eventually some were found to be in the possession of a practising consultant psychiatrist member, and after sharing a couple of documents, he sought the Group's permission in 2003 before divulging any more. The LCG would not allow my attendance at its meeting to explain the purpose of the research and, despite the apparent support of the member in possession of the papers, the group refused the request on privacy grounds.

Oral history interviews and the small number of accessed papers were therefore used to provide as full a picture as possible of the role of the LCG. This frustrating experience, however, illuminated the nature of the LCG: it had succeeded in controlling information which dated back to 1968, across generations, despite the fact that it was not centrally held and its existence was possibly even unknown among other members before the meeting at which access was discussed. The consultants' strong 'group' characteristics shown in their sense of solidarity meant that an individual member did not feel able to act autonomously but needed corporate permission to proceed, and its secrecy showed a clear boundary to the outside world.

Origins and purpose

The oldest grouping was that of the London Clinics consultants, brought together by the Ministry of Health[6] when the new Clinics were set up. In the early 1970s, the consultants broke away from the Department of Health, and after a period of rotating between hospital sites, moved to the Home Office in 1977 at the invitation of Bing Spear, then Deputy Chief Inspector at the Drugs Branch.[7,8] Informal, exclusively clinical meetings ran in parallel, at least for some of this period, initially hosted at St Bartholomew's Hospital and held during the evenings.[9] Both the formal and informal groupings are referred to collectively as the 'London Consultants Group', as they shared a great deal of business, membership and perceived identity. Indeed members of the groups give conflicting accounts as to which meetings were which.[10]

The formal LCG was initially chaired and minuted by Dr Alex Baker, a medical civil servant.[11] According to one member, 'the idea was there that policies would be determined'.[12] The LCG's meetings at the

Ministry of Health followed on from central government's direct rule over the Clinics, a result of the funding arrangements for London teaching hospitals that bypassed the Regional Health Authorities, although this later changed. Meetings seem to have been held initially about every month, attended also by Department of Health and Social Security (DHSS) and Home Office Drugs Branch staff.[13] Without civil servants the informal meetings were more relaxed and 'a place where we all got together and sort of said "Well my patients are worse than yours." '[14] Membership of both were restricted to the London area, where most of the Clinics were situated in 1968. Later on, groups were formed, with Home Office Inspectorate involvement, to encompass treatment centres in the rest of Britain corresponding roughly with the Inspectorate's own regional divisions.[15]

Following the new influx of doctors treating addiction outside the Clinics, Dr Dally started AIDA in November 1981 with encouragement from Bing Spear. He initially allowed the group to convene at the Home Office,[16] until they were forced to move to Dr Dally's premises in Devonshire Place.[17] Unlike the formal LCG, this state involvement was much more discrete and tentative. As an 'impartial' civil servant, Spear had to be careful of being seen to endorse any group, particularly one which aroused the hostility of the London consultants. He later defended his role, saying that while the London consultants were hostile to the Drugs Inspectorate's contact with AIDA and the private prescribers, 'That contact was perfectly consistent with our long-established policy of keeping in contact with anyone working in the drug dependence field. It did not imply approval, or disapproval, of the clinical judgement of those concerned.'[18]

The self-proclaimed 'independent' of the title referred to doctors working outside the Clinics, both as NHS GPs and private psychiatrists. This was a significant distinction as doctors inside and outside the Clinics had different prescribing privileges. Outside doctors had been discouraged from involvement in the treatment of addiction until official policy changed after the 1982 *Treatment and Rehabilitation* report (see Chapter 2). It was a division perceived by doctors on both sides of the divide, and *Treatment and Rehabilitation*, to which four NHS psychiatrists contributed, addressed prescribing in these terms.[19]

The AIP also came together with encouragement from the state, but this time in its local form through Kensington and Chelsea and Westminster Health Authority (KCWHA) and in response to grass-roots concerns about diverted pharmaceuticals, which was coincidental to a move by private doctors and their allies to defend themselves in the face of adverse publicity.[20,21] Tabloid newspaper the *News of the World*'s sting on private psychiatrist Dr Dzjkovsky, where a reporter had posed as a patient to test the ease with which he could obtain drugs, encouraged a seedy image of private prescribers.[22,23] The original meeting from which it developed had been called by KCWHA and organised by Siwan Lloyd Hayward, then project manager for Westminster Drug Action Team, representing local services and interests, with encouragement from public health director Dr Sally Hargreaves. It included representatives invited from the Home Office and Department of Health, medical and pharmaceutical professional regulatory bodies, local councillors, police officers, social services representatives and residents of areas concerned about street dealing.[24] Dr Matthew Johnson, part NHS GP and part private prescriber, Michael Audreson, Practice Manager at the private Hanway Clinic, and Gary Sutton, a private patient and activist, were instrumental in forming the AIP but it proved to be the shortest lived and the least influential of the three groups.

Initially an association of private doctors, it widened its borders to encompass other clinicians involved in treating drug users 'since common ground in respect to the treatment of patients should form the criteria for membership'.[25] Its multi-disciplinary membership was reflected in a name change in 1998, becoming 'The Association of Independent Practitioners in the Treatment of Substance Misuse'. This expansion would suggest that although most of their activities were aimed at the regulation of private prescribing, an important defining factor was not only the sector in which members worked but, like AIDA before it, a belief in the value of maintenance prescribing to addicts and in their distinctiveness from the NHS Clinics.[26–28] Its declared aims were to:

(a) define, describe and support prescribing outside traditional NHS Drug Dependency Unit standards;
(b) set up positive communication between practices;
(c) develop self-defence policy in case of problems with GMC or Home Office.[29]

The multi-disciplinary but above all non-NHS Clinic membership also reflected a reality of private practice in the 1990s, which far from a discrete sector, had many ties with the voluntary and statutory sectors. Matthew Johnson, for instance, received referrals to his private practice from an NHS DDU and prior to that, from a voluntary sector project in north London.[30]

AIDA's purpose in the beginning was to raise standards among doctors working outside the Clinics, share information and campaign for policy changes. At its first meeting at the Home Office, the group's stated self-regulatory intention was 'to define accepted standards of practice'.[31] In the following year, it developed rules for its members and then expelled a Dr Rai for apparently failing to follow them.[32] As discussed in Chapter 4, there was considerable concern at the time that *Diconal* was being prescribed in tablet form and crushed and injected with dangerous results.[33] AIDA decided that *Diconal* should be prescribed only in exceptional circumstance. Dr Rai, who was accused of persistently prescribing the drug, protested that he had changed nearly all his *Diconal* patients over to other medications and felt 'rather hurt' as he had 'tried at all times to comply with the wishes of the Association',[34] but the Association seems not to have relented. That same year Dr Dally, its president, admitted that she was still prescribing the drug herself: 'For a long time I have pressed and campaigned for the prescribing of *Diconal* to be restricted. I treat only four *Diconal* addicts. One of them will be off this drug within a week or two. One has never had any other drug than *Diconal* and is therefore a "pure" *Diconal* addict.'[35] She was then disciplined by the GMC for her *Diconal* prescribing. Similar inconsistencies happened within the London Clinics, where some long-term patients continued to receive heroin and injectable methadone in the 1980s after the Clinics had moved away from such prescribing and condemned it in the 1984 *Guidelines*, but without the regulatory consequences that afflicted Dr Dally.

AIDA's ambitious and confident aspirations to shape events reflected the sense in the 1980s that this was a crucial point of transition. Some of the biggest changes in the scale and nature of England's drug use were happening, and particularly in opiate consumption. Heroin smoking, previously unseen in England, became popular, a huge black market developed where previously most users had taken the overflow from doctors' prescriptions, much larger numbers of users were turning up for treatment, and the Clinics were inundated. Significant numbers of doctors outside the Clinics were being drawn into the field for the first time since the 1960s, and in the mid-1980s HIV/AIDS emerged to change the picture further (see Chapter 1).

The LCG's original aims included mutual support and sharing useful information. Thomas Bewley recalled them as 'rather jolly meetings', and that DDU psychiatrists tended to be rather isolated, especially within psychiatry where they were already looked down on by other doctors.[36] This need to discuss the work they were undertaking resulted partly from the sense of experimentation and uncertainty detectable in the early years of the Clinics. Before treatment allegiances solidified, many types of prescribing were tried out. Prior to the Mitcheson Hartnoll clinical trial of oral methadone and injectable heroin[37] (see Chapter 3), a nurse at Dr Mitcheson's Clinic recalled

> We actually prescribed all kinds of drugs; it was almost like a kind of oriental bazaar. People would come from far and wide to the Clinic and they would actually say 'We are using three or four or five grains[38] of heroin' and the doctor would say 'No, no, no, that's too much, but we give you three, and if you can't manage, we give you some physeptone [methadone] ampoules, and if you can't go to sleep, we give you some barbiturates and if you can't wake up, we give you some amphetamines.' It was this kind of bargaining at the beginning, until Dr Mitcheson had the research going.[39]

Dr Bewley, who attended LCG meetings until his retirement in 1988, concurred: 'No one had the faintest idea of what they were doing [at the Clinics] and were all expected to solve the problem of drug dependence, so it helped to swap notes.'[40] Practical matters and innovations were put forward. A formulation of methadone mixture including blackcurrant syrup developed by Dale Beckett was discussed, as the blackcurrant would apparently show up in patients' urine to show whether they had consumed the drug as prescribed.[41,42] This later became the accepted basis of the oral methadone formulation.[43] The formal LCG meetings also gave an opportunity to provide feedback to civil servants about forthcoming legislation.[44,45] At one meeting, the Home Office's representative Mr Beedle 'agreed to consider the possibility of the early introduction of regulations under the 1967 Dangerous Drugs Act restricting the prescribing of methadone'.[46]

Compared with AIDA, the AIP had narrower ambitions, perhaps reflecting the more stable policy period of the mid-1990s. Furthermore, with its origins in the complaints of a number of regulatory bodies and concerned parties, it was based partly in a need for self-defence, meaning that it was more concerned with changing the behaviour of its members and communicating these changes than with wider drugs policies. Like AIDA, it also expelled members, partly because it needed a sanction by which to enforce its standards, but also because some members felt tainted by association with particular prescribers and threatened to withdraw if this action was not taken.[47]

Both the AIP and AIDA meetings admitted to poor standards among some of their members, with particular concerns about financial motivation. Discussing the possible new clinical guidelines from the

Department of Health at a meeting of the AIP in 1997, Dr Brewer conceded, 'the essence of the problem was due to some private doctors increasing their patient loads to increase their financial gain and this could be the main motivation for treatment'.[48] Dr Beckett recalled his concerns about some of his fellow private prescribers in the 1980s: 'I remember that I used to go up to London to [Ann Dally's] consulting room, to her flat at the top of her house every so often and meet with other doctors who were prescribing because it was a worry really. A lot of doctors didn't seem to be doing right by their patients – giving them enormous prescriptions, extraordinary, and raking the money in – it was ghastly.[49]

Organisation and structure

Working in independent small businesses, private doctors could be characterised as 'low grid' and 'low group' entrepreneurs (see Figure 6.1). This term refers to their forms of organisation, ways of working and belief systems but does not imply that profit was their main motive. In Cultural Theory terms, such individuals have been described as practising in a context dominated by competitive conditions, control over other people and individual autonomy, and where the definitions and boundaries through which they related to the world were weakly drawn and flexible.[50] As independent contractors, GPs shared some of these characteristics, but with their greater dependence upon a single client, the NHS, had less autonomy and were less competitive among themselves. AIDA's structure and experiences reflected these characteristics.

Throughout its lifetime, AIDA's forward thrust was powered by a single charismatic leader, rather than shared among equally motivated members. Dr Dally, as AIDA's first and only president, seems to have chaired most of the meetings and undertaken the largest part of the work arising, such as writing letters and policy documents. When her involvement and interest in the drugs field came to an end, there was no one to replace her. Dr Dally's ability to network, a characteristic of the successful entrepreneur, was largely behind AIDA's high profile, given its small membership and the hostility of the opposition. As well as contacts in the media, she had Oxford University connections with one member, Dr Susan Openshaw, and also with Mrs Thatcher, then Prime Minister, which she used to lobby for her own and AIDA's cases. She took an opportunistic approach to recruitment, inviting diverse people to join who shared her viewpoint, such as American academic Arnold

Trebach. As might be expected of a small organisation with limited resources, the secretariat was provided in-house by Dr Dally.

Its purpose also largely reflected Dr Dally's concerns. At the outset she gained considerable publicity for AIDA, with announcements on the BBC's Six O'Clock news and in a number of medical journals. She campaigned for changes to the Clinics' approach to prescribing, changes to the law, and for greater involvement in prescribing by doctors outside the Clinics. Its degree of formal organisation seems to have diminished, starting out with both a president and a secretary and a Working Group to draft its own guidelines. As time went on there was a merging of her own personal difficulties with the GMC with the wider concerns of her colleagues, partly as a result of her view that the personal was political in this case,[51] and partly because of her dominant role in what became a relatively unstructured organisation. By the end of its life in 1987 to 1988, Dr Colin Brewer described the Association as a support group for Dr Dally 'in her time of trial'.[52] He recollected no regulatory role or intentions in the late 1980s. By 1987 AIDA was unstructured with no committee. Eventually the Association folded 'for lack of interest'.[53]

Like AIDA, the AIP was a small organisation whose secretariat was provided by the administrative staff of one of the member organisations – the Hanway Clinic – and had little internal structure. The Association also functioned as the first register of doctors providing private treatment to drug users. This was achieved by Dr Johnson, Michael Audreson and Gary Sutton pooling their knowledge for invitations to the first meeting, talking to other private prescribers and to pharmacists.[54] They limited their scope to doctors with significant involvement in the area, rather than those with one or two drug using patients on the grounds that they would have been harder to find and probably less committed.[55] The lack of any such register prior to this reflected, in contrast to the NHS, the lack of a central bureaucracy employing doctors in this field and little economic impetus to group together. Having said this, data held by the Ministry of Health or DHSS on the existence of NHS drug clinics was more than once found to be inaccurate.[56] In addition, some private doctors had been wary about publicising their services for fear of being inundated by addict patients seeking prescriptions.[57]

Just as the entrepreneurial character of the 'independent' doctors revealed itself in the organisations they developed, so it was with the London Consultants. These doctors worked within the hierarchy of hospital medicine with the NHS as employer. They shared a strong sense

of identity as a group, perhaps partly engendered by their lowly status within psychiatry and medicine as a whole, and consequently drew
themselves inside a boundary against outsiders. They were also motivated by a desire to keep at bay government involvement from what
they saw as clinical decisions.

While not competing for fees or patients, they were to some extent
rivals for prestige and resources, but steps were taken to minimise
competitive behaviour in the interests of the group. For instance, the
rivalry between Philip Connell, Director of the Maudsley Hospital's
drug treatment unit, and Griffith Edwards, Director of the Addiction
Research Unit at the Institute of Psychiatry, was not allowed to prevent the group from sending a letter of congratulations when Edwards
was awarded an academic chair.[58] Furthermore, to ensure that no one
individual gained too much power over the group, the role of chair
revolved between members. Likewise after a period during which the
meetings had been held at the workplaces of members in the 1970s,
the Home Office was taken up as a permanent venue when offered, as
it was geographically central and 'neutral ground', not being the base
of any particular consultant.[59] The allocation of tasks according to rank
is also typical of hierarchy.[60] According to Dr Martin Mitcheson, the
most senior members or 'elder statesmen' Drs Connell and Bewley were
deputed to visit Dr Dally to discuss with her the group's concerns about
her practice.[61]

The fight to guard clinical autonomy emerged early on. According to
Dr Bewley, the Department of Health's Dr Alex Baker erroneously minuted the first meeting to say that they had all agreed to reduce their
prescribed doses of heroin: 'There was general acceptance that complete
uniformity in prescribing practice was impossible. It was agreed however
that, as a general guide, each clinic should seek to reduce progressively
the total quantity of heroin prescribed.'[62] Thomas Bewley objected, seeing this as Dr Baker putting in his own opinion which had not been
discussed.[63] Although Philip Connell expressed the same view a year
after and this became the group's own policy,[64] it may have been the
government source of the proposal that made it unacceptable in the
minutes of that first meeting. A struggle for control developed between
the Department of Health and the doctors, with the doctors victorious. Early in the 1970s they wrote a letter telling the Department of
Health that they wished to break away and form their own independent group.[65] John Mack, consultant in Hackney DDU and the longest
serving member of the group, expressed the consultants' determination for independence from government. 'At the very early meetings

we wanted to make it quite clear that they were our meetings, they were not Department of Health meetings. We were happy for the senior medical officer from the Department of Health to be there and Bing Spear to be there, but we wanted them to be our meetings, not official meetings.'[66]

Some years later, in the late 1970s or early 1980s, another small battle took place, as one member recalled, 'There was a bit of an awkward scene from one time ... Dorothy Black was Senior Medical Officer [at the Department of Health] and she was from outside London and she came to the meeting and she tried to take it over and she had to be quite rudely told that it was not her meeting, that she was the Senior Medical Officer at the Department of Health being invited to our meeting.'[67] At the same time, the informal LCG meetings took place in the evenings at medical venues without any civil servants, but included discussion of points raised at the formal DHSS meetings, and agreed points to feed back to the DHSS secretariat.[68]

Membership and beliefs

A problem afflicting both AIDA and the AIP was falling membership. Dr Dally described the first AIDA meeting as well attended, with sixteen doctors in all including, rather surprisingly, one hospital consultant psychiatrist, some private psychiatrists, NHS GPs, Bing Spear and Ian Heaton from the Home Office Drugs Inspectorate and Dr Dorothy Black from the DHSS.[69] But after that, numbers seem to have diminished, with attendance at meetings generally only three or four.[70,71] Its president complained in December 1982, 'I can't say that we are inundated with applications for membership.'[72] The cost of membership in 1986 was £25 a year. At this stage they were meeting 'every few weeks'.[73]

The AIP also started with wider enthusiasm than it managed to maintain, and in a bid to set standards of practice, expelled the two members thought to be most problematic. Matthew Johnson explained, 'Well first of all we did throw a couple of people out of the group, who shall remain nameless. So ... trouble is after that what was left wasn't a very big group and ... then people stopped coming and ... the group that was eventually left were people who had been abiding, who had been well within the guidelines that had been set anyway, originally.'[74]

Key differences revealed by the three groups' policies were not only their content but their ability to reach and implement agreements. Although the focus of AIDA varied over the years, Dr Dally's essential message remained the same. Her own and AIDA's professed policies were that:

(a) 'the proper person to treat an addict [was] his or her own GP or a doctor to whom that GP has referred him or her';[75]
(b) long-term prescribing was necessary to allow stable addicts to maintain a law-abiding lifestyle, and the policy of the Clinics offering only short-term prescribing and detoxification was forcing such addicts onto the black market to obtain a drug supply;[76]
(c) NHS treatment was the ideal, but that until it was provided in a manner more acceptable to patients, private practice would continue to be necessary;[77] (AIDA tried unsuccessfully to set up a non-profit clinic for those unable to afford private fees, and applied for funding under the Central Funding Initiative of 1983.)
(d) prescribing injectable methadone had therapeutic value;
(e) *Diconal* should only be prescribed in exceptional cases;[78]
(f) drug treatment should take into account the role played by the criminal black market in the drug supply.[79]

Unlike AIDA, the AIP considered private treatment valuable in itself, rather than simply as a supplement where the NHS was inadequate, wishing to emphasise the fact that it did not burden the taxpayer and provided choice.[80] Like AIDA, it supported maintenance prescribing, including injectable methadone.

One of the spurs to the AIP's project in self-regulation was the anger and fear felt by residents of Sherland Road in west London towards the open street market in diverted prescribed drugs around Maguire's chemist[81] supplied by private prescribers. Diversion, it was admitted by the Association, was 'a reality' and 'the Achilles heal [sic] for private prescribing'.[82] This prompted a number of the AIP's policies, including the use of test dosages, where a patient took their first dose under the doctor's observation to ensure it was safe and that they were not asking for more drugs than they needed in order to sell on.[83] The AIP tried to produce guidelines laying out its policies. Although these allowed considerable latitude, such as not setting restrictions for how often patients should pick up their prescribed drugs which was an area of contention throughout the period, the Association was unable to reach agreement and they remained, like AIDA's guidelines, forever in draft form.[84]

The LCG's policies extended over the entire period under study, and changed during that time. Due to the LCG denying the author access to meeting documents, only a selection of policies are discussed. During the late 1970s and early 1980s, there was a spectrum of opinion among members over appropriate prescribing, but the majority seemed to follow Dr Bewley's view that prescribing should be standardised to present a united front to patients. So the aim of the meeting was to try to synchronise practice.[85] There was disagreement over the prescribing of heroin and injectable drugs, but, as shown in chapters 2 and 3, peer pressure successfully reduced heroin and injectable methadone prescribing in the 1970s and replaced it with oral methadone. The move away from prescribing injectable drugs and methadone in particular, which took place in the late 1970s, was foreshadowed in an LCG meeting as far back as December 1969: 'Although it was agreed that methadone undoubtedly had some value in the treatment of heroin addiction there was disagreement about the extent to which it should be used, particularly in its injectable form.'[86]

While AIDA and the AIP both attempted to develop guidelines with the aim of improving existing poor practice, the LCG seems to have produced guidelines early on with the aim of co-ordination. A document entitled 'Practical matters relating to the treatment of drug-dependent patients' was submitted by Dr Connell to a meeting in September 1969 at the Department of Health.[87] This may have been the same document minuted as 'Principles of Treatment', about which 'the meeting accepted the value of a document on the lines of Dr Connell's paper' but 'there was some disagreement with specific points'. A further draft was to be presented to a future meeting incorporating amendments from the group.[88] It is not clear whether the document published in the journal *Addiction* in 1991 is the same one mentioned in the minutes, and if so, whether this was the revised version or the original. At a meeting in 1969, the LCG determined to protect their prescribing expertise declaring that, 'The prescribing of methadone to addicts by general practitioners was unanimously condemned and it was agreed that a letter expressing this view to the medical press signed on behalf of those present as representing a body of authoritative medical opinion on drug dependence, might help to curb the practice.'[89]

On a number of policy developments Drs Connell and Bewley dominated. As with Dr Dally, this may have partly rested on personal qualities – one member remarked 'Philip Connell could walk into this room right now and take over; that was Connell's style',[90] but a stronger source of power was their extensive involvement in medico-political life and the prestigious offices it yielded. Both were, at various times, members of the GMC and special advisors to the Chief Medical Officer; Dr Bewley becoming president of the Royal College of Psychiatrists and

Dr Connell chairing the influential Advisory Council on the Misuse of Drugs (ACMD).

The class system also seems to have been a strong influence in hospital medicine, at least in the 1970s. Selection for consultant posts at St Bartholomew's consisted not only of the formal interview, but was preceded by a 'trial by sherry', essentially an informal drinks party at which the candidates were assessed,[91] with social experience and skill clearly playing a role. This hint at upper class, Oxbridgian preferences, is maintained by the note from an early informal LCG meeting, stating that an initial payment of five shillings would be charged 'to cover the cost of sherry for the meetings'.[92] The Home Office meetings, by contrast, offered a more bourgeois tea and biscuits.

Internal and external influence

Both the AIP and AIDA attempted to produce clinical guidelines for their own organisations and neither got beyond the drafting stage because of an inability to agree.[93,94] According to the AIDA minutes, 'There was a good deal of discussion about the way in which patients should be assessed.' Dr Poncia, a private doctor 'felt that routine urine testing in all cases might be counter-productive. This highlighted a certain amount of disagreement about the clinical management of patient [sic] in which members agreed to differ'.[95] AIDA and the AIP's members may have chosen an 'independent' path outside the NHS hospital setting in part because they did not like working to corporate policies, guidelines and protocols,[96] and this may have influenced their unwillingness to agree on guidelines. Furthermore, being to some extent in competition with each other for patients, they also had something to lose by working to the same patterns, particularly with the market-sensitive issues of cost and dosage.

Unlike the three editions of the official Department of Health guidelines,[97-99] which chose to limit the discussion of controversial topics, such as heroin, stimulant and injectable prescribing, the AIP's draft guidelines focused on regulating existing controversial practices. While the Department of Health's documents were aimed partly at encouraging the participation of doctors not already treating drug users, the AIP's document was intended to tackle the aspects of treatment that were gaining negative publicity and regulatory attention for doctors already involved. Other aspects of private treatment, such as the use of clonidine and lofexidine in detoxification used by some of the private doctors were not in dispute and so went unmentioned.[100] Matthew Johnson also proposed that a prescribing limit of 200 mg methadone be implemented between members, and 'There was a consensus of agreement that there could be an agreed upper limit, which all doctors involved in the group could work to,'[101] but it was never implemented. The AIP 'disintegrated in sort of people disagreeing too much we couldn't get proper consensus going'.[102]

Although both the AIP and AIDA finally collapsed, Dr Johnson and Dr Dally both felt that they had had a positive impact on their members. Dr Dally claimed 'I am quite sure that some of the less ethical doctors have improved their ways as a result of membership.'[103] Dr Johnson, while frustrated with the process of disagreement and fragmentation, believed that by bringing people together, prescribed dose levels had been reduced, even among the two suspended doctors, although neither claim can be verified without further data.[104]

AIDA focused not only on changing practice among doctors but also the views of the public and government. Ann Dally engaged in much media activity, including radio and television. She wrote letters to government and, with Dr Dale Beckett and Dr Tessa Hare, presented evidence to several committees. Articulate, intelligent and a good networker, Dr Dally had an impressive ability to gain the attention of influential individuals. She also benefited in this from promoting a consistent message in which she held no doubts. She was friends with and had some support from the editor of *The Lancet*, Ian Munro, whose journal published a strongly written attack on the GMC's handling of Dr Dally's first disciplinary hearing.[105] He later indicated his support for Dr Dally's criticisms of NHS drug treatment stating that 'the inflexibility of the present system is deplorable'.[106] However, one gets the impression that some of the medical press, rather than take sides, merely enjoyed provoking debate on its pages. Dr Munro, in a letter to Dr Dally, described how his 'misgivings ... from the rooted belief that treatment in this area must be separated from any kind of private practice',[107] had prompted him to write a leader article which took an opposing line to Dr Dally. He invited her to respond, saying, 'This *Lancet* contains a leader on drug addiction. The line it takes will hardly meet with your unreserved approval. Why not offer me a letter for publication? Anyway, let's hope we can stir up some debate on this shambles.'[108] Of the wider media, Dr Dally complained 'I talk to many reporters. Only about 1 in 10 writes down anything that seems remotely like what I said.'[109]

Perhaps AIDA's biggest opportunity came when Prime Minister Margaret Thatcher, with whom Dr Dally had been at Somerville College, Oxford, invited her to discuss the drug problem at Number 10. Dr Dally visited Margaret Thatcher at Downing Street early in 1983 but failed to convince her of her position. Indeed she speculated later that this high profile meeting may have encouraged her opponents to construct the GMC case against her.[110] Although she received some warmly worded letters from the Prime Minister, Mrs Thatcher made no interventions on AIDA's behalf, writing, 'I know very well how deeply you feel about this. But I hope you will understand when I say that I think it would be wrong for me to comment on the disagreement between yourself and the Department of Health and Social Security. I have read every word of your letter – but I cannot judge who is right.'[111]

Despite her impressive networking skills, Dally's access and conviction did not bring with it influence. AIDA's position, although of interest to the media, expressed dissatisfaction with the wider regulatory system of drug control in Britain, a view that, in the 1980s ruled it out of serious consideration. While AIDA maintained a high profile for several years, it also failed to gain influence in formal policy-making bodies. Chapter 3 showed how, although invited onto the 1984 DHSS good practice Guidelines Working Group, AIDA's representatives Dally and Beckett were sidelined and outmanoeuvred by Bewley, Connell and the Secretariat. While AIDA was pleased to see the proposed extension of licensing dropped by government, this was most likely due to Drugs Branch objections and ministerial concerns that the arrangement would require extra spending and risk alienating GPs rather than because of AIDA's opposition.

AIDA was also invited to give oral evidence to the Social Services Committee's enquiry into the misuse of drugs in 1985. Here again, the role of the London NHS psychiatrists may have neutralised any positive impact, for the Committee's special advisor was Dr Martin Mitcheson, consultant in charge of University College Hospital's DDU and a prominent member of the LCG. In lockstep with the London consultants, the Social Services Committee's report pressed for an extension of licensing to all injectable opiates, for licences to continue to be restricted to doctors working in or under the supervision of a consultant in a Clinic, and for particular attention to be paid to restricting private doctors.[112]

During the AIP's short life, the group made only one attempt to influence an outside policy-making body and failed. The Department of Health's Guidelines Working Group, whose final output would be published in 1999, had already drawn up its membership and started meetings when the AIP met for the first time and proposed nominating (uninvited) Dr Colin Brewer as its representative.[113] However, it seems unlikely, had they emerged earlier, that they would have gained access since relations were poor between Dr Brewer and the chairman John Strang. Strang had skilfully made a nod in the direction of private sector representation by inviting Dr David Curson from the Priory Clinics which used only drug-free detoxification. Had the AIP established itself prior to the start of the Clinical Guidelines Working Group, perhaps a representative would have been chosen for the sake of appearance, as occurred in 1984 with AIDA. However, the need to be seen to be consulting private prescribers was considerably less in the 1990s when they represented a far smaller proportion of the expanded drug treatment world, and neither were they the main advocates of maintenance prescribing. Indeed by 2007, the fourth Guidelines Working Group included no private doctors at all.[114] In general the AIP received and courted little media attention, although its work was highlighted in Radio Four's 'File on Four' broadcast in 1997 when the Association was attempting to write its own guidelines. The programme gave some recognition to its intentions to self-regulate but was largely critical of private prescribers.[115]

The LCG's external influence was stronger than the other two groups but it did not always achieve the policy changes it sought. The changes which followed the second Brain Report, on which certain London psychiatrists who became Clinic consultants had been highly influential, had succeeded in handing over heroin prescribing and the medical treatment of drug users to the Clinics in 1968, and they clung keenly to their prescribing privileges over the next three decades. Their attempts, however, to extend their monopoly over particular formulations and other drugs of dependence were unsuccessful. This disappointed aspiration was first recorded at an informal LCG meeting in 1969, when, 'It was regretted that the suggestion that prescription of all dependency producing drugs to known addicts should be limited to treatment centres had been dropped by the Department of Health on grounds of cash.'[116] After this came moves to own the prescribing of methadone or other opiates. Although gaining support from many influential bodies, such as the Social Security Committee, the ACMD and the Department of Health's 1999 Clinical Guidelines Working Group, the consultants remained unsuccessful here also (see Chapter 7).

Aside from these disappointments, Connell and Bewley achieved some successes in protecting the interests of the NHS Clinic psychiatrists and extending their discipline over outside doctors. Both were members of the ACMD's *Treatment and Rehabilitation* Working Group, and its report recommended a number of curbs on the prescribing of doctors outside the Clinics.[117] Although not all of these were implemented, the proposed good practice guidelines for doctors did become a reality. Dr Connell was awarded chairmanship of the committee responsible for the first guidelines, and John Strang, the most senior addiction clinician at London's Maudsley Hospital, chaired the next three published in 1991, 1999 and 2007. Specific measures to protect the privileged position of NHS Clinic psychiatrists were included in both the 1984 clinical guidelines and the 1999 re-write,[118,119] and both were used in disciplinary cases against doctors working outside the Clinics before both the GMC and the Home Office's Drugs Tribunals.[120]

The successes of the LCG in influencing outside policies were not simply through decisions taken at its meetings but resulted from the involvement of many of its members in other important bodies. It was therefore less the meetings themselves, than the perceived sense of being a group with shared interests that could be promulgated in different arenas. While the AIP and AIDA members rarely met outside their own Association meetings and were 'weak group' in Cultural Theory terms, several of the London consultants were colleagues at the Royal College of Psychiatrists, responsible for the postgraduate training of psychiatrists and influential over a range of psychiatric policy. The Royal College of Psychiatrists was also empowered to nominate a member to the GMC, first sending Dr Philip Connell in 1979, and then his replacement Dr Bewley in 1981. Several LCG members were also on the ACMD, parliamentary committees and Department of Health Working groups.

Although individual consultants occasionally appeared in the media, both formal and informal incarnations of the LCG kept low public profiles. For instance, in 1969 a formal Department of Health organised meeting agreed to write to the medical press to express its 'unanimous' condemnation of GPs prescribing methadone.[121] The letter was published in the *British Medical Journal* on 16th May 1970, although Bing Spear later accused the Clinics of misrepresentation, and that excessive quantities of injectable methadone available for sale on the streets originated principally from the Clinics themselves.[122]

The LCG, in contrast with the two Associations, did not generally publicise their own rules as a group. For instance, it was not until 1991 that the 1968–69 guidelines were published by Philip Connell, and then only as a document of historical interest.[123] While the London consultants did not observe all their own rules or agreements this quieter approach gave them fewer 'hostages to fortune' than AIDA, whose publicly announced rule on the prescribing of *Diconal* was to trip up Dr Dally during her first GMC hearing in 1983.[124] Furthermore, where rules were influenced by the London consultants, such as the

1984 *Guidelines*, they still managed to fend off outside intervention or scrutiny and accomplished the feat of setting rules for other doctors to which they themselves did not have to adhere.

Dr Dally's approach to the media differed from Drs Connell and Bewley, reflecting the Cultural Theory characteristics of each and the traditions of policy-making in the drugs field. An entrepreneurial net-worker, Dally wished to garner support from any quarter that could help her case. She did not perceive a strong barrier around those inside or outside her group. The LCG, which reflected the hierarchical structure of hospital medicine, by contrast, was restricted to doctors practising in the Clinics. With a strong boundary between insiders and outsiders, members did not usually publicise their views to the general media, preferring to keep the debate on drugs within the medical realm of conferences and journals. Dr Connell and Dr Bewley rarely appeared in the general media to rebuff Dr Dally's accusations and criticisms.

Policy-making in the drugs field in the 1960s and 1970s was carried out behind the scenes in private by accommodation between experts and civil servants.[125] This held true for the LCG and it may have been the public nature of Ann Dally's attacks on the Clinics that so embittered the London consultants as much as the content of the attacks themselves. Discussing in public what the London consultants saw as matters for private, or medical only, discussion broke their code of private policy-making, risking involving patients and the wider public.

Relations between the groups

Just as relations between private prescribers and the London Clinics were generally antagonistic during much of the 1980s, this too was the case between AIDA and the LCG. The minutes of the AIP revealed no references to the London consultants as a group; their concerns were directed more at self-regulation and self-defence. It has not been possible to trace whether the LCG was aware of the AIP during the latter's brief existence, or what its reaction might have been to its activities. Although there was no chronological overlap between AIDA and the AIP, some members (and even a draft of AIDA's guidelines) were shared between the two.

AIDA supposedly wanted closer co-operation between the doctors outside and inside the Clinics. In response to John Strang's 'Personal View' article in the *British Medical Journal*,[126] Dr Dally wrote a letter to its editor for publication, explaining that, 'Clinics can also provide what independent doctors usually cannot provide, for example, group decisions in patient treatment and group psychotherapy.' She concluded in conciliatory tone, 'It is vital that the Clinic and independent doctor co-operate with each other. Failure to do so along the lines suggested by Dr Strang can only harm the patients.'[127] In a letter written to *The Lancet* for publication, as AIDA's 'Founder and Organiser' she suggested a number of measures to help the drug treatment situation, first of which was 'concerted efforts of doctors and others who work in the field to co-operate and not to descend to slanging matches about how awful, stupid, indifferent, greedy or wicked the other groups are', a standard she was not herself able to maintain.[128] She wrote to the newspapers criticising the approaches of the Clinics, even naming individual doctors such as Connell.[129] The constant criticism aimed at the Clinics by AIDA suggested that a spirit of co-operation was not being fostered. AIDA's draft guidelines themselves opened with a number of volleys aimed at the Clinics, and several of the letters written for publication contained attacks too.[130]

In 1982 Dr Dally sought a meeting with the Clinic doctors but apparently with no success. Writing to Ted Hillier in the DHSS's drugs branch, on the suggestion of Bing Spear, to establish communication with the 'Clinic doctors', she complained, 'We have made a number of overtures to them but have met with no success. We would very much like representatives of this Association to meet with appropriate people to discuss matters of common interest.'[131] Consultant Martin Mitcheson did not recall these 'overtures' but took the view that they would have been resisted had they been received.[132]

Regulating other doctors

Both AIDA and the AIP's regulatory gazes were directed mainly at their own members but the LCG set the rules by which other doctors were regulated. Consultants with concerns about the practices of other doctors could take a number of paths. They might contact the Regional Health Authority whose advisors could visit a doctor, write to the individual doctor themselves, write to the GMC or mention their concerns to a Drugs Branch Inspector. This last approach could arise through an LCG meeting at the Home Office, or through a range of other contacts doctors had with the Home Office. Martin Mitcheson, for instance, used to visit Bing Spear in the course of his research at the Addiction Research Unit of the Institute of Psychiatry,[133] and doctors phoning the Drugs Branch to notify the Addicts Index could also discuss regulatory action.

Before the 1984 *Guidelines* had been published, over which the London consultants had been decisive, the Home Office Drugs Branch relied in part on advice and publications from the London consultants as to the appropriate practice standards they should enforce.[134] Furthermore, there were no cases of Home Office Tribunals being used against the London Clinic doctors for inappropriate prescribing. This was not because they all adhered to the *Guidelines* during the 1980s or 1990s; by the late 1990s and 2000s they were not even offering a uniform range of treatments or doses but rather because they were left to self-regulate. A long-serving inspector who worked in the Drugs Branch since the late 1980s described the Clinics' practice in the 1990s and 2000s,

> They all prescribed somewhat differently...there is a core of activities that are common to them all but then there are others; Strang would do injectables, [St] George's don't do injectables and things like that...there are slight differences. I think they tend to see that they should be held up as the model prescribers...Seeing as John Strang chairs most of these committees anyway, in a sense he should be doing what they said and other consultants accordingly.[135]

He described the process of regulation as follows:

> The London consultants have a quarterly meeting is held here [Home Office], which I've attended since 1987. And part of that is sharing of information about drug misuse, prescribing, and sorts of things like that. So if I'd have, which I didn't have, if I'd said I'd got concerns about a particular doctor who was a consultant, I'd have probably spoken to Hamid Ghodse initially [convener of London Consultants Group]. And he would have, or John Strang, or someone like that, and perhaps persuaded them or perhaps asked them to perhaps have a quiet word in their ear about what's going on. Certainly they always felt that they should be supportive to each other, and that if things, there was one of their doctors going out of line they should try and put them on the straight and narrow.[136]

The idea of doctors being 'supportive to each other' showed the shared sense of a group interest and identity that was weak in AIDA and the AIP. The inspector answered the question 'What if one of the Drugs Branch had concerns about a consultant's prescribing?' as follows:

> *Home Office Inspector:* Well, if they did, I never heard about it. There was this one occasion when I can think of where the other consultants had concerns about someone prescribing and that needed to look at the boundaries of their...

SM: So the consultants were self-regulating in that sense, they kept an
eye on each other.
Home Office Inspector: Yes.[137]

So here was the LCG informally reporting non-Clinic doctors to the
Home Office Drugs Inspectorate for regulatory investigation, apparently
immune from challenge by the Inspectorate, while concerns about their
own members were dealt with among themselves. The single exception to this reliance upon informal, internal regulation was the case of
Dr Kanagaratnam Sathananthan. He was not only an advocate of continued heroin prescribing on a maintenance basis after the Clinics had
changed their practice, but one of only three doctors ever licensed to
prescribe heroin privately. The GMC and the Home Office Inspectorate
became interested in his private and NHS prescribing respectively but
he retained both his freedom to practice and his Home Office heroin
licence. If it was Dr Connell, as has been suggested,[138] who was behind
these inquiries, it seems likely that Dr Sathananthan was treated differently by his colleagues because he was a private prescriber. Whether his
outsider status was also due to being originally from Sri Lanka is unclear,
although Iranian born Hamid Ghodse was very much an LCG insider.

Relations with the state

The role of the state in each organisation was different and acted not as
a monolithic entity with a single interest but part of a complex policy
network itself, reflecting and pursuing a number of different interests.
AIDA's encouragement from the Chief Inspector of the Drugs Branch,
represented Spear's support for doctors to be free to give injectable and
maintenance prescriptions and his concerns about the monopoly of
treatment provided by the Clinics. Spear contributed a factual section of
AIDA's draft guidelines describing the legal position of doctors prescribing for addicts and the notification procedure for the Addicts Index,[139]
but his views on treatment policy were very discreetly held while
employed. During his service he promoted his own preferences through
quietly supporting others behind the scenes, but after retirement he
began to campaign more openly. In 1987 he wrote to one of Dr Dally's
supporters, praising Dr John Marks, a dissenting NHS psychiatrist who
was a vocal proponent of heroin prescription at his Liverpool DDU:

> I am not too despondent as there are signs that a rethink is around the
> corner and a more flexible approach [to prescribing] may be adopted.

I think we should all do what we can to support those doctors, like Dr Marks in Liverpool, who are proposing this and I suggest, when the election is over, you should put your point to your local MP.[140]

In the memoirs published after his death, Bing Spear lamented,

With the benefit of hindsight there is no doubt that the treatment centre era was an unmitigated disaster, not because the basic idea was wrong but because of the way in which that idea was developed and implemented. What happened was that the moral high ground was seized by a small group within the medical establishment, and by psychiatrists in particular, who, over the years succeeded in imposing their own ethical and judgemental values on treatment policy. As a consequence there is now very little prescribing of heroin, or any injectable drug, to addicts.[141]

While discreetly encouraging AIDA, Mr Spear was also wary of becoming too closely involved, writing, 'I remember, after attending an AIDA meeting at which a very doubtful prescriber was present, noting that we should be careful in our dealings with the Association because it was by no means unlikely that some of those who applied for membership might in due course be regarded as candidates for Tribunal action.'[142] Dorothy Black, while Senior Medical Officer at the DHSS, attended the first AIDA meeting and commenting upon its draft guidelines. She too was careful to distance herself, writing to Dr Dally to correct the minuted description of herself as an 'observer', a status apparently restricted to civil servants attending major external meetings.[143]

Spear's support for AIDA did not represent ministers' direction of policy. Indeed the lack of a strong interest from politicians in the finer points of prescribing left civil servants to form their own policies within their wider brief. In this way, Spear worked to his own agenda, including support for prescribing heroin and alternatives to the Clinics. He knew almost everyone in the drug treatment field and formed alliances to push through his policies. For instance, when a complaint was made against Dr Sathananthan, he sought out researcher Cindy Fazey to assess his private clinic. As a sociologist, Fazey was an unusual choice, but known to be sympathetic to Dr Sathananthan's style of prescribing. According to Professor Fazey, it was Philip Connell who was behind the attacks on Dr Sathananthan. With his hatred of Connell, Spear interpreted this as a vendetta, motivated by personal dislike.[144] Fazey's report exonerated Sathananthan's prescribing. She also appeared as a witness for the defence at GMC hearings of both Dr Sathananthan and Dr Dally. Along with John Marks she formed part of an anti-London-Clinic, pro-maintenance faction. This is not to say that Dr Fazey's political perspective influenced her report on Dr Sathananthan. Professor John Strang, strongly against private prescribing, conducted an investigation into Dr Sathananthan's NHS clinic but also found no unprofessional conduct.[145,146]

The AIP had little contact with central government but had originated from a meeting with the local government and local statutory agencies. The Director of Public Health for Kensington and Chelsea and Westminster Health Authority, Dr Sally Hargreaves,[147] continued to liaise with members of the AIP about progress on self-regulation. This culminated in a joint project between the NHS and the private sector, where KCWHA funded the Hanway Clinic to provide a drugs clinic for homeless patients in Soho; an example of the blurring boundaries between public and private common in the 1990s.[148] However, there was little sense that government wished to consult these doctors on policy changes. Although Anthony Thorley, Senior Medical Officer at the Department of Health, stated that there would be a consultation on the draft clinical guidelines that were being revised, and that their input would be welcomed, the absence of any private prescribers on the committee itself reinforced the impression that this was mainly cosmetic.[149] In the event, there was no consultation on the draft guidelines, which were published in their finalised form in 1999. As the LCG's origins were tied to central government, its relationship has already been discussed above.

Unifying and divisive forces

Both AIDA and the AIP had diverse memberships but shared a belief in the value of maintenance prescribing and treatment outside the NHS Clinics. The working arrangements of the members were probably less important than their beliefs about drug control and supply. For instance, Dale Beckett had been an NHS Clinic consultant psychiatrist for many years before going private, but he believed in far more liberal access to drugs and free prescribing of heroin. NHS GP Diana Samways resigned from AIDA, protesting about its emphasis on maintenance prescribing, and the charging of fees, writing, 'I feel very concerned about the prescribing of drugs (and for money) to addicts, it seems to me that AIDA is a forum for the justification of this. I also felt that any other views on the treatment were heresy, and not for discussion at AIDA.'[150] As well as the treatment approaches that members of the AIP favoured, they were also drawn together by a sense of threat from the media and regulatory bodies.

In addition to the views that developed on prescribing outside the Clinics, many members of the LCG, such as Drs Willis, Bewley, Mitcheson, Ghodse and Connell, specifically opposed private prescribing where a doctor was paid directly by patient fee.[151–153] Where private prescribers were thought to be a problem, they were discussed and it was then for the Home Office to decide on whether to take action.[154] However, Martin Mitcheson recalled a deputation of consultants themselves being sent on behalf of the group to visit Ann Dally and express their concern. There was some uncertainty over this event however, as Dr Dally made no mention of this in her autobiography. What is more certain is that Dr Bewley reported Dr Dally to the GMC in 1984, although it decided to take no action.[155]

The LCG seems to have had the strongest sense of locality, which it drew around itself as a boundary to outsiders. Within London the members defined catchment areas for their patients and there was a belief that, at least in the 1970s and 1980s, the London scene was unique in scale and patterns of drug use. With this came a rather unreceptive attitude towards their peers working in the provinces, one of whom recalled, 'I can remember as a clinician coming down from Sheffield, in the late, probably about '79 and talking about *Diconal* and I can remember the London consultants looking at me as if I had no idea what I was talking about, because they'd never heard of the drug, because it wasn't being used in London whereas it was a major problem in the north of England.'[156] Equally, a member of the LCG explained, 'There was also the feeling that we had a lot happening between ourselves with our patients and most of the activity around treatment was around the centre of London anyway and quite frankly, I think some of us got rather fed up hearing someone like [consultant from outside London] telling us what type of tablets were popular with his ten addicts … We had a lot of things to talk about amongst ourselves.'[157] The AIP inevitably drew its membership from in and around London, as large scale private prescribing was virtually unknown outside the South East of England, and AIDA, which encouraged national membership from GPs, had a much wider spread. This may have made it more difficult for its members to meet, as they had further to travel.

An ability to trust each other gives the members of a group an advantage in working together as information can be shared openly. As might be expected from a group with a greater sense of shared purpose, a strong boundary drawn against outsiders and less direct competition between members, there was greater trust within the LCG than with the AIP, whose members were in competition with each other. At one AIP meeting, 'It was noted that clients can and do change doctors if they can access higher levels of prescribing even though they have managed well on lower doses. This was an issue of concern.'[158] At the AIP's meeting on 13th March 1997, those present discussed the possibility of forming a consortium to buy urine tests and detoxification units,[159] but this never came to fruition, probably because buying as a consortium would have revealed sensitive information such as which doctors were not using urine tests. A year earlier forms had been distributed in order to collect information from doctor members on their patient caseloads, fees, and other details. There was reportedly some reluctance to complete these partly for commercial reasons and also because one could work out a doctor's gross income by multiplying fees by numbers of patients.[160] In contrast John Mack described an implicit confidentiality of the meetings of the LCG; an understanding that you could speak freely in front of your colleagues and civil servants without risk of information leaking out.[161] The fact that an individual member did not feel free to share papers until the group had been formally consulted reinforced this sense of trust and basis for confidentiality. Unfortunately the data available on AIDA did not give a clear picture of the degree of trust between members.

The antagonistic position AIDA began to take towards the Clinics may have reduced AIDA's strength and appeal both within and outside its membership. While they were serving together on the DHSS's Medical Working Group, Ann Dally invited NHS GP Arthur Banks to join her Association. Dr Banks was well respected in the drugs field and had considerable experience in treating drug users, receiving praise for the booklet he had written on the subject.[162,163] Although in agreement with Dr Dally on many issues, both opposing the extension of licensing for prescribing doctors in 1984 and keen to attend a meeting of AIDA, he declined to join, explaining, 'There seem to be very widely divergent views in the drug treatment world, with clinics and independents and social-model workers often strongly condemning each others' policies. I am torn between the various views, or perhaps trying to remain neutral; I share many of the criticisms of the clinics but am not happy about being "independent" either.'[164] Having a member such as Dr Banks would not only have helped in terms of achieving external influence but he could also have helped train and advise AIDA's less experienced members. Dr Samways remarked in her letter of resignation from AIDA, 'I am sorry to hear the negative attitude AIDA members have to the Treatment Centres, and having worked in the St Bernard's Unit,

I am aware of the problems we all face.'[165] Dr Black, Senior Medical Officer at the DHSS, chided Dr Dally for her oppositional stance in AIDA's draft guidelines.[166]

The LCG suffered disagreements within its ranks too, with a range of opinion on, for instance, the prescribing of injectable drugs. Within the LCG the proponents of a particular point of view, such as Drs Connell and Bewley regarding the opposition to heroin and injectable methadone, were able to get their views adopted by most of the group and achieve a change in practice across the Clinics in the 1970s and 1980s. Once again, the importance of presenting a united front to outsiders (in this case patients and other doctors) rose above members' individualistic impulses. Greater variations in practice seem to have occurred later in the 1990s but insufficient evidence precludes further comment here.

A problem that seems to have occurred with both AIDA and the AIP was a feeling among some members that disreputable doctors were using the associations to gain personal legitimacy but had no intentions of changing their practice. A letter from an NHS GP and AIDA member early in the organisation's life expressed concern that 'AIDA might act as a front for potentially unscrupulous doctors wishing to benefit from prescribing privately for drug addicts.'[167] Dr Dally herself was concerned about this possibility from the outset. Writing to GP member and friend Susan Openshaw, she asked her for advice on 'what we should do with people who quite definitely are using it as a blanket of respectability and who are not attempting to keep up high standards of practice.'[168]

Matthew Johnson felt that this syndrome afflicted the AIP too and undermined the other doctors' willingness to lend support. The expulsion of two doctors might have helped this, but then, given that the organisation had no other sanctions to apply, the expelled doctors could continue to practice outside the association, which then had no influence on them at all.[169] Both were eventually struck off the medical register by the GMC. The private prescribers of the AIP seemed to divide into three groups: those attracting adverse publicity who wanted to improve their respectability through association; those with no regulatory difficulties, who wanted to 'keep their heads down' and perceived that they had nothing to gain from associating with less respectable doctors; and those who wanted to achieve change and improve the standing of private prescribing through group action. Doctors with no trouble from the regulatory authorities, such as Dale Beckett and Jeremy Bullock, did not attend the AIP, as they perceived no need to club together for protection. Being individualistic operators they did not take the view that to 'attack one of us is to attack all of us'. This inevitably reduced the number of 'respectable' members.[170] Furthermore, once the first two groups were not attending – those not interested, and those using it for their own purposes – the remaining doctors were few and it was a case of 'preaching to the converted'.[171]

Most of the London consultants already had respectability within government and among the public thanks to their positions within the NHS and 'establishment' organisations, although they complained of low status within psychiatry.[172] Promotion within the NHS relied upon being acceptable to one's peers and superiors, criteria missing from the private doctors, who could work in their own businesses regardless of selection procedures. With the exception of Dr Sathananthan in Croydon, members who were perhaps more unorthodox, and continued to prescribe heroin, such as James Willis, either left of their own accord or were subjected to other external pressures;[173] Dr Dale Beckett's NHS clinic was closed down reportedly because of the hospital's dislike of drug user patients.[174,175] The position and role of the LCG during the 1990s is less clear due to the inaccessibility of any documents from those meetings. Unlike Philip Connell in his day, the most senior consultant at the end of the century, Professor John Strang, rarely attended the meetings, and it seemed that a greater diversity of approaches to prescribing was tolerated inside the London Clinics, but determining the reasons for this would require greater access to source materials than is currently possible.

Conclusion

Comparing the AIP with AIDA points up the changed policy environment facing private prescribers in the 1990s. The AIP still attracted some interest from the Department of Health and Home Office, and psychiatrists and central government civil servants were still using opportunities for national policy-making to regulate private prescribers in the 1990s, as in the attempted extension of licensing in 1999–2000. Yet the origins of and response to the AIP showed the growth in significance of local policy-making in the drugs field and the marginalisation of private prescribers. It also reflected that London was no longer the dominating interest of drug policy as drug use and services had proliferated across the country.

A key weakness of both the AIP and AIDA was their lack of formal sanctions that could be applied to non-conforming members, other than expulsion, which left them unable to enforce their policies. Yet this could equally be said of the LCG, so what made the difference? The London Clinics, created by the Ministry of Health to address the problems identified in the second Brain Report,[176] already had a stronger relationship with the state than the private doctors. Its leaders embedded themselves further within establishment bodies, such as the GMC and Royal College of Psychiatrists, which they could then marshal against perceived outside threats. Furthermore, informal sanctions, such as face-to-face disapproval, would have been harder to avoid for the London consultants who encountered each other across a number of settings and so could have had greater impact, strengthening the group's power over individual members.

Strong leadership from the forceful Dr Dally was not enough to bind together a group of independent individualists working outside the hierarchical hospital system. They were unwilling to compromise their autonomy for longer-term gains and so failed to produce consensus guidelines, instead 'agreeing to disagree'. By contrast the LCG was willing in the 1970s and 1980s to make the sacrifices in individual autonomy required by its leadership to increase its corporate autonomy. This can be explained through the wider institutional power bases of the dominant Drs Connell and Bewley beyond their personal qualities, the hierarchical nature of hospital medicine, and the multi-stranded relationships within the LCG.

Max Gluckman, in his analysis of feuding societies in Africa and their settlement mechanisms, identified different allegiances across a number of settings as the root of social cohesion: a feud with someone in one arena threatened that relationship across several settings and therefore more was at stake and there was a greater interest in settling the dispute.[177] The London consultants encountered each other in the Advisory Council on the Misuse of Drugs, at the Royal College of Psychiatrists, at the Society for the Study of Addiction, and on working parties. AIP and AIDA members rarely encountered each other in different occupational settings, and felt they had less to lose by staying true to their own preferences. Subsequent events suggest they were mistaken.

7
Guidelines and the Licensing Question

Introduction

Unorganised and leaderless, the eight years between the Association of Independent Doctors in Addiction's passing and the Association of Independent Prescribers beginning saw private prescribers lose any foothold in the policy community. Yet conversely, this was a time when the message of their former leaders had been taken into the heart of the policy-making. Those who had advocated harm reduction for many years but had lacked political legitimacy were able to advance their policy objectives under the new threat of AIDS. With the risk of the HIV virus spreading not only among drug users through shared injecting but from them into the general population through unprotected sex, perceptions about the most urgent goal of treatment altered. Abstinence made way for the goal of reducing the harms associated with continuing drug use.[1] The UK government officially endorsed this pragmatic policy approach known as 'harm reduction' in 1988. Oral methadone and needle exchanges became central to this response. At the end of the 20th century, treatment goals shifted again, with concern about acquisitive crime perpetrated by addicts overtaking HIV/AIDS as the urgent prevention priority of the moment. Yet oral methadone maintenance endured to serve this goal too with support from inside the Clinics as well as outside.

It was into this altered landscape that the third edition of the 'Orange Book' or the clinical guidelines on drug misuse emerged in April 1999.[2,3] It was the biggest, the most heavily referenced, with the longest production period and the largest Working Group of the three guidelines so far. Just as in 1984, the Working Group had also been asked to make a number of unpublished recommendations to ministers covering a system for licensing doctors to prescribe controlled drugs for the treatment of drug misuse and this time also regarding training clinicians, monitoring prescribing practice and improving the supervision of consumption of prescribed controlled drugs.[4] A memo written by the Working Group's secretariat shows the intentions of the new licensing, referring to

> ...a particular problem with inappropriate methadone prescribing leading to diversion onto the illicit market in a small number of private practices, particularly where practitioners work alone and where a majority of their work involves substitute prescribing for heroin and other drug misusers. The Working Group will be making particular recommendations to the Department of Health and the Home Office Drugs Inspectorate about how all practitioners who prescribe inappropriately might be more effectively monitored and controlled.[5]

The licensing proposals seem not to have been aimed primarily at private prescribers but would have addressed many of their practices that Professor Strang and the Drugs Inspectorate under Alan Macfarlane found unacceptable. Private prescribers were a much smaller and less significant issue by the end of the century than they had been in the 1980s. The 1999 *Guidelines* repeated policy concerns dating back to the late 1960s and 1980s but also reflected the altered treatment environment of the late 1990s. Past continuities could be seen in the attempt to regulate prescribing by private doctors and NHS GPs, particularly for injectable and other opioid scripts.

The major changes from the previous two versions of the guidelines reflected wider political changes, developments in the country's drug misuse, in treatment policies and services and in the nature of clinical guidelines themselves. These included: the relentlessly increasing scale of UK drug use; a growth in services and expertise beyond the traditional London centres; the increasing participation of GPs and non-medical professionals in treating drug misuse; the Department of Health's more developed role in treatment policy; the emergence of HIV and the policy responses around it, particularly the new international orthodoxy of methadone maintenance; the policy aim of a 'primary care-led' NHS changes in key policy personnel at the Home Office and Department of Health; the 'evidence-based medicine' movement; and a run of medical scandals resulting in calls for tighter regulation of the profession. This third edition of the *Guidelines* grew out of a number of these changes.

Nineteen eighty-four had seen the introduction of general management into the NHS and the overt encouragement of local decision-making. Paradoxically, the government saw this devolution as requiring extensive central co-ordination and encouragement through its provision of a multitude of guidelines, directives and circulars. This could also be seen in the drugs field a little earlier, where efforts to develop local services, often in the voluntary sector, through the Central Funding Initiative (1983–89) were orchestrated by Whitehall (see Chapter 1). As well as stimulating the voluntary sector, the government provided modest incentives for people to take up private health insurance and ended Labour's opposition to private beds in NHS hospitals. Over the 1980s, the number of private hospitals providing abstinence-oriented treatment for drug-dependent patients grew considerably.[6]

A boost to the trend for clinical guidelines came with the introduction of the internal market into the NHS from 1989. Without the levers of a true market, all kinds of government mechanisms were developed to try to make healthcare more measurable and comparable for contracting decisions between GP purchasers and hospital or community service providers. Questions about what constituted good quality care fuelled a new guidelines industry. Clinical audit, introduced in 1990 with generous Department of Health funding, was a new tool for measuring the outcomes of treatment, and for changing treatment where it was considered deficient. Doctors' leaders participated grudgingly and practitioners were obliged by government to do so.[7] In order to define good treatment, guidelines were needed here too.

To meet this demand, academics and professional medical bodies developed expertise on the development of guidelines, encouraged by Department of Health funding.[8] This, along with the movement for 'evidence-based medicine', partially arising from the medical profession, led to greater formalisation of the production of guidelines and an insistence that they be based upon formal research studies. The evidence-based medicine movement helped to legitimise the use of guidelines within the profession and although most elements of the internal market were dropped by the Labour government, the revival of managerialism as a driver of change found favour in their continued use.

In the 1990s, new standards were set for clinical guidelines by the Department of Health requiring greater formal use of research evidence and external review. Guidelines issued in the early stages of the evidence-based medicine movement or before had made little reference to published research evidence: the 1984 *Guidelines*, probably the first official guidelines document across the UK health services, contained no references to scientific studies, only reports, textbooks or reference sources such as the British National Formulary. The 1991 edition referenced fewer than five research studies,[9,10] but the 1999 edition contained almost one hundred research references.[11]

Accompanying these changes were a number of moves that strengthened the position of GPs within the health service, such as fundholding, that gave primary care doctors greater control over their budgets and enlarged their scope to provide additional services. Fundholding also changed the balance of power between hospital consultants and their GP 'customers'.[12] There were disincentives for GPs to send their patients to hospital, leading to the emergence of 'GP specialists' expert in the treatment of a particular patient group or condition. For the treatment of chronic diseases, GPs were encouraged to enter into 'shared care' arrangements with hospitals. These could follow a wide range of models but the essential idea was that specialists and GPs would plan a patient's care together, explicitly sharing out various aspects of the work between them.[13]

The development of consumerism both outside and inside the NHS was increasingly important over the whole period. Standing out against this trend, the 1999 *Guidelines* were more typical of NHS drug treatment policy-making that showed minimal consumer input before the 21st century. Like other expert committees in the drugs field, such as the Advisory Council on the Misuse of Drugs (ACMD), the Working Group's membership lacked any patients, although there were two ex-users on a sub-group. In the 21st century as patient (or 'service user') groups became more active both in the UK and internationally in the drugs field, and were encouraged by some parts of government, their representatives were included on more official bodies. The National Treatment Agency for Substance Misuse established by government in 2001 made a strong stand on involving patients in decision-making about their care and the planning and delivery of services.[14] In 2007, the Working Group to revise the 1999 edition of the guidelines included both service users and carers.[15] Users had finally breached the prescribing citadel but the extent of their influence needs further exploration.

The 1999 guidelines

In 1996, the Department of Health had published the 'Effectiveness Review', which, as well as commissioning new research, attempted to review all the evidence on treatment and services for drug users in the largest such undertaking at that point.[16] Its introduction made clear that the Department of Health was already planning new guidelines on the clinical management of drug misuse to replace the 1991 edition. This may have been to reflect the newly reviewed literature and perhaps also in response to the Review's own recommendation about the need to restrict prescribing of injectable drugs to particular doctors, which had originated with John Strang.[17,18] Strang was Britain's most senior drug dependence psychiatrist, Director of the National Addiction Centre at the Maudsley Hospital and Institute of Psychiatry and had chaired the 1991 Guidelines Working Group. He found support for the new guidelines and licensing in Anthony Thorley, a fellow Maudsley-trained psychiatrist and Senior Medical Officer at the Department of Health, and Drugs Branch Chief Inspector Alan Macfarlane.

Unlike previous Clinical Guidelines Working Groups, the 1999 edition's membership was not exclusively medical, acknowledging 'the active role now played by other professionals'.[19] Nurses, pharmacists, social workers, psychologists and the voluntary sector had been widely involved in drug treatment well before the first guidelines were written, so why had this not been reflected until the late 1990s? Back in 1982 the *Treatment and Rehabilitation* Working Group had recommended that the first guidelines be produced by an all-medical group, feeling unable to comment on prescribing issues itself. With the wider rise in consumerism and the questioning of the bases of many kinds of authority and privilege since the 1960s, the areas in which doctors could claim medical autonomy, free from the influence of outsiders, were under constant pressure over this period. Among expert committees in drugs policy, the ACMD became less medically dominated and the Effectiveness Review's Task Force (1994–96) had been overwhelmingly drawn from the non-medical world. Yet doctors had successfully defended prescribing as their sole preserve, remaining the only professionals in addiction treatment able to sign prescriptions for controlled drugs. Despite medical attempts to hold off increased non-medical influence, sometimes successfully, prescribing eventually succumbed to at least a public acknowledgement of non-medical input in 1996. The last bastion from encroachment, from that of the patients themselves, was held but Roger Howard, Chief Executive of the Standing Conference on Drug Abuse, followed in the traditions of the voluntary sector in the drugs policy community, speaking on behalf of the absent users.[20,21]

The Working Group also included for the first time representation from Northern Ireland, a representative from the General Medical Council and two public health doctors. Strang's chairmanship of the 1999

Guidelines Working Group indicated that addiction psychiatrists had managed to maintain their position as the expert authorities in the drug treatment field. They also made up the largest specialism with seven of its eighteen members. All four GPs from England, Scotland and Wales had special experience in drug problems. Clare Gerada was to become a part-time senior policy advisor at the Department of Health and with Michael Farrell took on the drafting of the *Guidelines* from 1998 after Anthony Thorley's departure.[22]

The Association of Independent Practitioners in the Treatment of Substance Misuse, representing private prescribers, had hoped that private addiction psychiatrist Colin Brewer could join the Working Group,[23] but made the suggestion too late after the group had been set up.[24] Dr Brewer ran the Stapleford Clinic, a large private prescribing practice that also carried out rapid opiate detoxification under sedation/anaesthesia, the practice of which Professor Strang was publicly critical.[25] (Dr Brewer was struck off the medical register in 2006 after one of his patients died following a home detoxification.) The AIP commented that 'if private prescribers were left out of the policy and decision-making in respect of the Guidelines then it would not be viewed as a collaborative effort'.[26] The chairman *had* in fact included private sector representation but he was not a private prescriber: David Curson was employed by The Priory hospitals, whose practice lay outside the private prescribing controversy as it did not involve substitute prescribing on an outpatient fee-paying basis. In-patient treatment in private hospitals and residential facilities was associated with the abstinence-based Minnesota Model, also known as '12-step' and familiar through Alcoholics Anonymous and Narcotics Anonymous.[27] A member of the secretariat described Dr Curson as 'the acceptable face of private practice' who was in favour of 'getting the rogues in Harley Street'.[28] David Curson therefore gave the Working Group a voice from private medicine while avoiding internal opposition from private prescribers, against whom John Strang, like Philip Connell his predecessor in the chair, had long been active.[29,30]

While Dr Brewer would have been an unlikely choice for the chairman to make, the likelihood of the chairman or Department of Health feeling compelled to invite a private prescriber onto the committee suggested a certain naiveté about the selection process, and perhaps an over-estimation of their own importance during the 1990s. In contrast with 1984, when the AIDA was a prominent organisation, the position of private prescribers in November 1996 when the letters of invitation were sent out had been further weakened by their lack of any representative body. The 2007 Guidelines Working Group, which formed after Dr Curson's death, included no private sector representation at all.[31]

With the Working Group selected, the business of discussing treatments and arrangements for their delivery began. Minutes of the 1999 Working Group's second meeting recorded:

> There was general agreement that the primary role of the doctor is not to ensure that individuals become drug-free – that is a moral issue – but to reduce the harm to individuals. However, where abstinence is essential to, or an efficient means of, reducing harm, that will be one of the goals of treatment. This message will inform the drafting of the Guidelines.[32]

'General agreement' may still have allowed some room for dissent. Some disagreement on these principles, in particular methadone maintenance prescribing, came from Dr Diane Patterson, Chair of the Northern Ireland Committee on Drug Misuse.[33] By the late 1990s, methadone maintenance had become much more widely accepted in the treatment policy community in Britain and many other countries. Although still arousing controversy on the 1999 Guidelines Working Group,[34] with the accumulation of strong research evidence and the support of those in influential positions, including the chairman, opposition proved ineffective.

A range of views could also be found regarding the extent to which the demands of public health or the individual patient were seen as paramount in prescribing decisions, a long-running tension in drugs policy across the UK. For instance, Dr Laurence Gruer, a consultant in public health medicine with Greater Glasgow Health Board, favoured indefinite supervised consumption of methadone by patients to protect others from the risks from diverted supplies.[35] This involved patients taking their prescribed drugs under the observation of a pharmacist or doctor to make sure they weren't giving or selling the drugs to anyone else. Chris(tine) Ford, a west London NHS GP and passionate advocate for the rights of drug users, who had been described as an 'NHS private prescriber', thought there should be no such stipulations, commenting, 'If you keep people on supervised consumption forever then they aren't allowed to move on or grow in any way. If you treat them like a child they behave like a child.'[36] Gruer himself agreed that the primary care side showed more of a sense of direct engagement with individuals, whereas the psychiatrically oriented members took a more intellectual approach.[37]

Probably the largest gulf existed between doctors inside the Working Group and those GPs outside who argued that treating drug problems lay outside their obligatory workload (core general medical services) and should be separately remunerated as a specialist activity.[38] Since 1996 this had become a topic of disagreement between the Department of Health and the British Medical Association's General Medical Services Committee, the GPs' main trade union.[39] The split this caused between 'experts' and 'ordinary' GPs (or their representatives from the General Medical Services Committee) had erupted on the British Medical Association's Working Party on Drug Misuse between 1995 and 1997 and produced almost complete paralysis for a portion of its fraught existence.[40] John Strang, an approachable man with an unconfrontational attitude to committee discourse, had found the British Medical Association's Working Party a jarring experience which may have determined him to choose GPs for his own Working Group for their expertise and enthusiasm rather than for their representativeness.[41,42] The first Guidelines Working Group had brought inside the opposition but then ignored its views.[43] Membership of the third Working Group not only represented the greater degree of consensus of its time but was also chosen for its ability to work together productively.[44]

As well as the Working Group itself, there were a number of subgroups charged with examining particular issues. Some of these, such as the private prescribing sub-group, were made up of existing members and secretariat or observers.[45] Others, like the injectable prescribing subgroup brought in outsiders including, for the first time, some patient representatives, two ex-users who were 'adamant against injectables' after experience of such prescribing, according to one member. They had been chosen by Duncan Raistrick, the sub-group's chairman, described as 'not a keen lover of injectables' himself.[46] In addition to the members were a number of observers, including Alan Macfarlane and, reflecting the proliferation of drugs agencies within central government, observers from the Central Drugs Coordination Unit.

Like the 1984 document, a key aim of the 1999 *Guidelines* was to allocate appropriate activities to different doctors as a basis for extending licensing and for disciplinary action. To do this, the 1999 edition introduced a new category, the 'specialised generalist' in-between the 'generalist' and the 'specialist'. This super-GP was not restricted to the drugs field but reflected the increased power, status and domain of general practice that had accompanied the flow of resources into primary care in the 1990s. The three categories were differentiated by experience, the proportion of their patients who needed treatment for drug problems, levels of training that they should receive and provide, and the requisite degree of autonomy or collaboration with others.

Criminal justice or public health concerns such as preventing diversion of drugs onto the illegal market were included as a treatment goal, as well as reducing 'the need for criminal activity to finance drug misuse'.[47] The lack of patient representation and influence on the Working Group was reflected in statements such as, 'Due notice should be given of a reduction regime', suggesting that change should not be imposed suddenly and without warning on patients but neither was the patient's agreement necessary.[48] The chapter on assessing patients' needs and situations repeated the 1991 *Guidelines'* warning that in private practice the doctor should 'establish that the patient is able to pay for treatment through legitimate means'. Despite the belief repeated in official documents since the second Brain Report,[49] that prescribing was only part of an overall approach to treatment and rehabilitation requiring psychological and other input, prescribing remained the focus here, taking up four of the seven chapters. Only one page was devoted to 'broader approaches to psychosocial support and treatment'.[50]

Prescribing remained the most controversial area, perhaps the one seen by addiction psychiatrists and policy-makers as having the potential to cause the greatest harm; opiates, the mainstay of years of prescribing debates, occupied the most space. The most significant prescribing change from previous editions of the guidelines was the endorsement of methadone maintenance as an activity suitable for primary care but tied to much strengthened and more specific recommendations for daily supervised consumption. The 1999 *Guidelines* put great emphasis on the proven efficacy of methadone maintenance but still expressed caution about the ability of ordinary GPs to prescribe any substitute drug unsupervised. Rather than bringing greater autonomy to the GP, the approval of methadone maintenance for primary care offered greater restrictions. The advice on amphetamine substitution was also a departure, conceding cautiously, 'There may be a limited place for the prescription of dexamphetamine sulphate 5 mg (five) in the treatment of amphetamine misuse.'[51]

The revival of international interest in heroin prescription, partly influenced by the positive results from a rigorous clinical trial in Switzerland,[52] may have prompted the first ever appearance of a section on this topic. Strangely no evidence was cited in the single paragraph which concluded, 'With the availability of injectable methadone, there is very little clinical indication for prescribed diamorphine.'[53] A similar section appeared on injectable prescribing, describing it as a specialist-only activity.[54] The tone of these *Guidelines* did not make the tasks described appear easy or straightforward as the first edition had, scattered as they were with numerous cautions on risks, pitfalls and safety precautions.

Although the 1999 edition was the most densely referenced of the series, and 'relied substantially on the major undertaking of the Task Force to review the evidence base for services for drug misusers', they were not the dramatic departure from the past. Many previous reports published in the 1980s and 1990s from expert committees in the drugs field had relied heavily upon the authority of their contributors, such as the ACMD's *Treatment and Rehabilitation*, or the first two editions of the guidelines, and contained very few references to published research evidence. In 1999, despite a considerable increase in publication on clinical addiction research in the UK and internationally, the *Guidelines'* introduction, under the heading, 'Evidence-based Guidelines', stated that they were 'primarily based on evidence obtained from expert committee reports and the clinical experience of respected authorities'.[55] Indeed, the section on maintenance prescribing, although referencing research reviews, looked to a quotation from an ACMD report for support, a document which contained only 36 references, several of which were policy documents and other ACMD publications.[56] Once again, it seemed, drug treatment policy was to be determined by 'respected authorities' albeit with some extra research backing.

The 1984 *Guidelines* had a number of clear-cut aims in the minds of the key movers behind them: the strengthening of the position of the Clinic psychiatrists and the authority of their model of treatment and the control of prescribing by doctors working outside the Clinics, most specifically private practitioners, through greater regulation. Prescribing outside the Clinics remained a concern on the 1999 Guidelines Working Group, addressed through its licensing recommendations.[57] Unlike the 1968 heroin and cocaine licensing system where doctors' suitability for licensed prescribing was based upon their specialty and location of work, something it was hard for GPs to change, under the new system a doctor who was able to gain sufficient extra training and experience could potentially qualify for a specialist licence.

One important theme of the 1999 *Guidelines* themselves and the Working Group meetings was defining primary and specialist treatment more clearly. While helping to promote shared care arrangements by removing ambiguities, showing GPs what was involved in this work,[58] these definitions also had a disciplinary function. In his evidence against Adrian Garfoot at that doctor's Home Office Misuse of Drugs Tribunal in

1994, John Strang spent some time wrestling with the appropriate distinctions between specialist and generalist addiction services, and this may have inspired him to seek firmer, formal definitions.[59] Marking clearly what was suitable prescribing for primary care would also prevent what was perceived as undesirable prescribing in the first place and make disciplining those who stepped outside the definition easier, either through GMC hearings or through the withdrawal of one of the new licences.

In spite of the more consensual nature of the 1999 Working Group, a last minute disagreement almost upset the whole process. With echoes of Dr Dally and colleagues' threat to produce a minority report, three of the GPs, Chris Ford, Judy Bury and William Clee, wrote to the secretariat threatening to remove their names from the final document. They protested that the draft produced by the secretariat after the final meeting of the group on 16th March 1998, from which Drs Ford and Clee had been absent, was radically different to previous versions. An examination of the drafts prior to and following the final Working Group meeting did reveal substantial changes both in structure and content.[60,61] Methadone maintenance was suddenly given much greater emphasis than any other intervention and many of the chapters, such as 'Young People and Drugs', 'Pregnancy and Neonatal Care' and 'Managing Drug Misuse Emergencies' had shrunk and been relegated to annexes. Other controversial topics, such as stimulant and injectable prescribing, were altered to sound less positive. But minutes of the final meeting recorded that these changes had in fact been suggested by the Working Group,[62] rather than effected through the secretariat conspiracy suggested by the GPs' letters.[63,64] Perhaps in the absence of some of the group's more liberal individuals who opposed greater regulation and controls, the other members took advantage to press through their own preferences. In fact the secretariat took on board some of the complaints made about these changes to the satisfaction of the GPs[65] who then agreed to endorse the document with consensus achieved more democratically inside the group than at the 1984 Working Group.

This was not the last hurdle, however. Before they were finalised, a small scandal broke out over the leak of minutes of a Working Group meeting to *Druglink*, the trade journal of the Institute for the Study of Drug Dependence, in an apparent attempt to whip up opposition to the licensing recommendations and get them modified. The response to this leak from the chairman was reportedly quite tolerant.[66] The Department of Health was aware that 'the culprit was a member of the independent working group ... on the grounds that the article quotes directly from the minutes of the last meeting of the group which were sent only to members.'[67] It is not clear whether the civil servant who wrote this knew or suspected which member was responsible, but if he did no action was taken.[68]

The main opposition to methadone maintenance arose from Belfast psychiatrist Diane Patterson. This was the first time that the guidelines had included Northern Ireland, where there were neither methadone prescription nor official needle exchanges.[69] While Northern Ireland had been happy to copy the first guidelines almost to the letter in their own edition, the move towards harm reduction in the intervening years had been resisted. Dr Patterson claimed that the emphasis on substitute prescribing for opiate addiction 'would place Doctors in Northern Ireland in an impossible position', and consequently she would be advising the Chief Medical Officer of Northern Ireland not to adopt the *Guidelines* pending the final publication.[70] This disagreement pointed up geographical differences in prescribing traditions, which one member considered a chief source of divergence across the Working Group.[71]

Dr Patterson's stand over the document seemed to have been prompted by the peculiar placing of a section entitled 'Methadone maintenance – the evidence' in the introduction to the document circulated after the last meeting. The text appeared before the discussion of any other aspect of treatment giving it an unnatural prominence.[72] The medical secretariat, Claire Gerada and Michael Farrell, tried to respond to these issues by arranging a meeting with the chief critics, Diana Patterson, Chris Ford, William Clee and possibly Judy Bury the following month.[73] Although placated with regard to the *Guidelines*, Ford and Clee remained unhappy about the licensing proposals and both signed a letter to *Druglink* opposing them. They saw these as specifically aimed at curbing 'the prescribing habits of a few private doctors in London' and being harmful to all drug users trying to access treatment.[74]

Although suspicious of private healthcare in general, Chris Ford echoed many of Ann Dally's arguments. Both expressed concerns about treatment for patients suffering withdrawal symptoms during assessment – the period when a patient had presented for treatment and the appropriate course of action was being decided.[75,76] Both were critical of the NHS Clinics and saw private treatment not as an ideal but as legitimately revealing shortcomings in existing NHS provision.[77–79] After the final meeting of the Working Group which had endorsed an extension of licensing, Chris Ford wrote to the chairman and secretariat in words that could have been written by Dally herself asking, 'What is going to happen to many drug users being provided services by the private sector, and perhaps many GPs who do not prescribe like the local specialist services? Many users are managed in general practice and the private sector because the NHS can't or won't provide the care they want or require.'[80] They also shared concerns about the dignity of patients in treatment, expressing greater trust in them than many of their colleagues.[81,82] Ford commented of her patients, 'If you believe people, they tend to tell you the truth.'[83]

Although termed 'guidelines' both the 1984 and 1999 editions were intended to be used as a tool for medical discipline rather than simply as a suggested approach.[84] The draft proposals from the Home Office and Department of Health for an extension of licensing was presented as a system of statutory control to enforce the 1999 *Guidelines*,[85] but as in 1984, the attempt failed. The 1984 edition had hoped that 'these guidelines would help to identify those cases where prescribing practices might be regarded as irresponsible.'[86] The 1991 edition made no mention of any regulatory role, but the 1999 *Guidelines* gave a stern warning that although they had 'no defined legal position ... any doctor not fulfilling the standards and quality of care in the appropriate treatment of drug misusers that are set out in these Clinical Guidelines, will have this taken into account if, for any reason, consideration of their performance in this clinical area is undertaken'.[87]

Like the 1984 edition, the 1999 one was intended to control private prescribing but as a much smaller part of its wider concern to define and regulate the appropriate practices of primary and secondary services with a view to both encouraging and controlling treatment outside the Clinics. While the issue of private prescribing played a part in the 1999 group's considerations, particularly in the questions of licensing and the prescribing of injectable opiates, the prominence of this almost exclusively south-eastern phenomenon in the wider drug treatment scene had diminished with the expansion of NHS services across the country and the participation of many more GPs treating drug misuse. The 1999 *Guidelines* were in fact intended not just for the guidance or discipline of doctors but also to reshape services. While the 1984 *Guidelines* were primarily aimed at doctors working outside the Clinics, the Inspectorate and the GMC, the 1999 *Guidelines* were additionally targeted at the bodies responsible for medical training and resourcing, including the Department of Health. They made recommendations, such as those around supervised consumption of methadone, which had spending implications, and expected government to respond accordingly. In this respect the 1999 Working Group was given a wider remit than its forebears.

Extending licensing

The concept of extending the licensing system introduced in 1968 had been around as long as the original system itself. As discussed in the previous chapter, a Department of Health meeting of the London Clinic psychiatrists the following year had proposed that *all* dependency producing drugs to known addicts, not just heroin and cocaine, should be removed from GPs and limited to the Clinics. The Department had rejected the proposal as too expensive.[88] The author was refused access to minutes of the London Clinic consultants' group meetings so it has not been possible to find out whether it attempted to get licensing extended during the 1970s. We *do* know that the idea was revived in 1980 and pushed forward in the ACMD's *Treatment and Rehabilitation* report. This time the measure was aimed more at controlling private doctors than GPs. In 1984 a vote on the issue by the first Clinical Guidelines Working Group found a majority in favour but the government delayed taking action. A year later, in 1985, the Social Services Committee, whose remit was to scrutinise the Department of Health and Social Services (DHSS), looked into drug misuse, treatment and rehabilitation, and made similar recommendations. These were to extend licensing to cover all injectable opiates and restrict licences to 'doctors working in, or under the direct supervision of, a consultant or equivalent in a Clinic'.[89]

The Social Services Committee had been advised by Dr Martin Mitcheson from the University College Hospital Drug Dependency Unit. He was a strong advocate of oral methadone prescription, a member of the London Consultants Group and an opponent of private prescribing where the doctor was directly paid by patient fees.[90] The Social Services Committee expressed great respect for the recent reports of the ACMD,[91,92] and seemed to have taken its line on prescribing regulation from these. At this stage, the government was yet to come to a decision about the Guidelines Working Group's proposals to extend licensing and the Social Services Committee urged it to do so. It did, resolving that it was unnecessary and possibly likely to deter GPs from treating drug users.[93,94]

The licensing recommendations of 1969, 1980, 1982, 1984 and probably 1985 came from senior Clinic psychiatrists seeking to contain and control prescribing, particularly of injectable or maintenance opiates, within the centralised state-funded Clinics, where addiction psychiatrists developed and tried to maintain their monopoly of expertise. A similar pattern emerged in the late 1990s, but for the first time boosted by strong civil service support. A new generation of politically active psychiatrists shared many of their predecessors' concerns, pushing for greater regulation of doctors working outside the Clinics. John Strang had trained at the Maudsley Hospital under Philip Connell, following him into a number of his policy and clinical posts. Whereas in the 1980s these medical men lacked the support of key civil servants at the Department of Health and Home Office, in 1999 the psychiatrists and administration were united. Alan Macfarlane was frustrated by the Home Office's existing mechanisms for controlling prescribing and the lack of enforceable standards. Referring to 'a current tribunal case where the doctor's position was that the needs of his particularly difficult patient group justified an approach which involved enormous quantities of drug cocktails', he complained of the inadequacy of the 1991 *Guidelines* for use in regulation: 'it is significant that the existing guidelines did not provide the Tribunal with a cut and dried benchmark'.[95,96] Macfarlane saw the new *Guidelines*, reinforced by extended licensing, as the solution to these problems. Anthony Thorley agreed with Macfarlane on this issue,[97,98] but Thorley's departure in 1998 altered the balance once again, weakening Macfarlane's position.

The idea of restricting injectable methadone prescription to particular doctors, however, had been in the mind of the chairman, Professor Strang, at least as far back as 1995 when attending the meetings of the Effectiveness Review.[99] He had written to its chairman, Reverend John Polkinghorne, about the possibility of extending the licensing system to cover injectable drugs. In his attached draft, which was included largely unchanged in the Task Force's report, written after discussion with the Reverend Polkinghorne,[100] he discussed concerns about injectable methadone, especially privately prescribed. Strang concluded that injectable prescribing should be limited to specialist doctors working in services 'with adequate multidisciplinary input, and with systems in place to safeguard against abuse of this service and to prevent diversion of the prescribed injectable drugs into the black market',[101] largely ruling out private prescribers.

This letter and draft were copied to Anthony Thorley at the Department of Health and the Effectiveness Review published a recommendation that 'The Department of Health should explore ways to ensure that injectable addictive drugs are only prescribed for drug addicts by doctors (including GPs) with appropriate training and expertise working with adequate multidisciplinary input and by specialist drug misuse services.'[102] The addition of GPs in parentheses hinted at a significant area of debate on the 1999 Guidelines Working Group: what prescribing should be carried out by GPs, particularly those with additional training and experience, several of whom were members of the Working Group? And should such GPs be able to prescribe in the same way as specialist psychiatrists in hospital-based settings? Right up to the final two drafts, the *Guidelines* switched between including only consultant psychiatrists as 'specialists' and allowing other doctors into this definition.

Although the 1999 Guidelines group was asked to make recommendations, the details of implementation were to be left to the Drugs Inspectorate. The Working Group specified that the Director of Public Health should be the countersigning officer for any licence application but it did not decide who should make the actual decision about the award of a licence, a large area of potential ambiguity that caused disquiet among doctors in the field.[103] The 1999 licensing proposals were intended by the Home Office and Department of Health to reduce diversion of pharmaceuticals and 'monitor good practice'.[104,105] John Strang's aim seems to have been similar, with particular interests in restricting injectable prescribing and enforcing the daily dispensing and supervised consumption of methadone. In his 1995 national survey of community pharmacies co-authored with Janie Sheridan, the authors had found widespread disregard for recommendations on dispensing and consumption of methadone in earlier clinical guidelines or ACMD advice, 'thus increasing known dangers of misuse and diversion to the black market'.[106]

Members of the AIP voiced concern at the prospect of licences for prescribing injectable methadone,[107] some private doctors fearing that it could be used to put an end to their practice. There were concerns that licences for injectables would be given only to psychiatrists working in Drug Dependency Units, as had been almost exclusively the case with heroin and cocaine licences since 1968. Daily dispensing and supervised consumption of methadone might also be unaffordable for private patients compared with a weekly or fortnightly pick-up from a pharmacy as they would have to meet the cost of the additional dispensing fees themselves.[108,109]

Anthony Thorley briefly considered the possibility that private doctors might be excluded from injectable licences 'for not being able to fulfil the specialist criteria'. He had sketched the characteristics of a specialist doctor in an earlier working party document as including (though not requiring) the 'capacity to provide specialist support to generalists in shared care setting', 'active use of specialist professional inputs from a multidisciplinary team', 'fast turn around access to pathology and drug testing services' and 'use or provision of specialist treatment techniques:

clinical psychology, counselling, etc.', all of which could be seen as outside the scope of most private doctors who often worked alone, outside the hospital setting and in isolation from medical colleagues.[110] However, most of the considerations around licensing submitted to and produced by the Working Group seemed to have assumed that at least some private prescribers would apply for and receive licences.[111] The 1999 *Guidelines* even saw a role for private doctors in NHS shared care arrangements, recommending, 'Where there are no local specialist services with which a shared care agreement can be developed, it is the responsibility of the health authority to ensure that appropriate services are in place. This might mean, for example, developing a shared care arrangement with a service in the independent or private sector.'[112] Such collaboration between private and NHS or voluntary sector services were not unheard of in the 1990s[113] and characterised a blurring of boundaries between public and private.

The sub-group of the 1999 Working Group set up to consider private doctors in the licensing system reported two possible options: an equivalent requirement for private and NHS prescribing, or a stricter licensing requirement for private prescribing, to include oral methadone because, 'This recognises the fact that as this is a unique situation where, in particular, payment is received for a prescription of a controlled drug with potential financial advantages from long-term prescribing, it therefore requires more comprehensive controls than the NHS.'[114] Ultimately, the Working Group recommended extra controls for private doctors to prescribe oral methadone not required of other doctors and they also endorsed the contemporary unwritten policy of Home Office Drugs Inspectorate restricting licences for cocaine, heroin and dipipanone to NHS prescribing.[115,116]

Once the principles of the licensing extension had been presented to ministers, the operational details were drawn up by the Inspectorate and sent to a range of organisations for consultation.[117] The psychiatrists had needed the support of civil servants to bring the licensing proposals to this point but it was not enough and they were never implemented. This was, as discussed in the previous chapter, due in part to greater self-regulation by the Royal College of General Practitioners who attempted to improve levels of training among their members in treating drug misuse, so fulfilling some of the Working Group's aspirations. The strongest factor was probably opposition from GPs keen to guard their clinical autonomy and, once again, fears in government that greater restrictions on prescribing would deter reluctant GPs from treating drug users.[118] Anthony Thorley's departure from the Department of Health in 1998, where he had been a strong advocate of the licensing system, also weakened the forward thrust of the policy. Michael Farrell, his replacement, was 'more of a clinician than a civil servant and less interested in the regulatory side'.[119]

Conclusions

If working by committee means that no one gets exactly what they want, the 1999 *Guidelines* were a case in point, representing successes, accepted compromises and failures for its members. The outlook for greater prescribing regulation had been improved by the turnover of civil servants. Key figures in the Department of Health and the Home Office, who had been influential in the first guidelines and licensing debates, had since moved on. Bing Spear had retired from the Home Office Drugs Inspectorate in 1986 (and died in 1995). Dr Dorothy Black, a consultant psychiatrist who had headed a drugs Clinic in Sheffield, left her post as senior medical officer responsible for drugs and alcohol at the DHSS at the end of the 1980s. Spear had been wary of handing too much control to the powerful London consultant psychiatrists and, along with Dorothy Black, was against extending licensing in 1984.[120,121] By 1997, when the idea was being reconsidered, Thorley and Macfarlane had taken their places. Unlike Spear, who was well known for his personal interest in the welfare of drug users and doctors' clinical autonomy outside the Clinics, Macfarlane was more concerned with controlling the drugs supply and preventing diversion. While getting closer than ever before to the goal of extended licensing its proponents failed to deal with the forces that stymied the attempts of the 1980s. However, they did make some gains: this pressure from the state, supported by some members of the profession, raised professional standards through the Royal College of General Practitioners' training certificate.

During its deliberations, the Working Group saw movement back and forth over what prescribing was considered suitable for primary care and the extent of any safeguards this required. The chief advocates of greater autonomy and less central control were GP members, showing more trust in their own judgement and their patients. Those directly employed by the state, such as the psychiatrists and public health physicians, put their trust in central government and in those doctors who, like themselves, had received more extensive formal training.

By contrast, Ireland successfully introduced licensing of GPs to prescribe methadone with its 1998 'Methadone Protocol', actually increasing GP participation.[122] The Irish licences required GPs to prescribe substitute drugs for opiate-dependent patients with the support of specialist services. While many circumstances bore striking resemblances to the English situation, the offer of a lucrative payment scheme to the licensed doctors may have provided the significant difference in gaining GPs' support.

Neither patients nor private prescribers were invited onto the 1999 Working Group, although some of its policies were directed at each, perhaps revealing that those selecting the membership saw them as requiring regulation rather than consultation. The absence of consumerist influences in this part of the NHS compared with, for instance the production of mental health service guidelines, was notable. The potential for conflict that these absences removed may have made the meetings run more smoothly but other fault lines emerged. The expansion of geographical representation to Northern Ireland for the first time emphasised the divergence of prescribing traditions across the UK, including differences between Scotland and the South East of England. Divisions between, on the one hand, public health and drug control issues, and on the other individual health concerns, emerged here once more, not as an inter-departmental split between the Home Office and Department of Health but between doctors themselves. Indeed, the two departments were allied in their attempts to control prescribing outside the Clinics, and those psychiatrists in favour also benefited from support from the new addition of public health doctors to the group.

Despite invoking 'evidence-based' medicine and referencing many more research studies than previously, the *Guidelines* were frank that their main sources were expert committees and respected authorities. In this they followed the tradition of the previous editions and of other expert committees in the drugs field. Evidence also seemed to be used unevenly. While the section on methadone maintenance, encouraged by the *Guidelines*, was given several references to research and evidence reviews, more sensitive topics such as heroin prescribing which they seemed to be trying to deter, had no references at all. Although methadone maintenance could boast a much larger evidence base, a large scale, well-publicised and respected clinical trial of heroin prescribing had been recently published but was not mentioned.

The policy-making process of the Working Group bore similarities to the old style behind-the-scenes doctor and civil servant negotiations of the 1960s and 1970s. However, the leak of confidential Working Party papers to the medical and drugs policy press went against this 'gentlemanly' code of private policy-making. Although it breached the agreed secrecy of the group, the leak may have been tolerated because it did not bring out these drug policy debates out into the general, public media but only to the drugs field or perhaps the chairman and secretariat had failed to identify the culprit.

The 1999 *Guidelines* appeared to endorse a more liberal approach to prescribing, for instance with its recommendations for methadone maintenance in primary care and cautious recognition of some role for amphetamine and injectable opiate prescribing; but they maintained a restrictive view of who was qualified to carry out this work. The guidelines and their accompanying recommendations fell more towards increasing the state regulation and control of those working outside the Clinics and their patients and reducing autonomy than the previous edition. However, the failure of the licensing proposals greatly weakened their intended impact. In contrast to previous guidelines, the larger number of participating GPs and their strengthened position both in drug treatment policy and in the NHS more broadly were reflected in the concept of the 'specialised generalist'. With additional experience and training, capable of treating more complex cases outside the Clinic system, this 'super-GP' represented increasing acceptance that doctors outside the hospitals could be specialists too.

The wider geographical spread of services and reduced prominence of the London prescribing scene in treatment policy debates strengthened voices from outside the metropolis and reduced the concentration on private prescribing issues, working both for and against the interests of private prescribers. Less attention was given to controlling their prescribing practices than in the 1980s but their scope for representation and participation in the policy process was also diminished. Instead, some of the GP members, although not necessarily in favour of private prescribing per se, shared many of their interests and fears: they acted as proxies for private prescribing on the Working Group, mitigating the centralising urges of other members.

Conclusion

This book began by asking how decisions about addiction treatment have been made in England and how conflicts between doctors affected national treatment policies. By now, it is clear that there are many answers to these questions. Who makes these decisions has changed, with doctors dominating in the 1960s and 1970s but gradually having to concede space to other professional groups and interests in the process. Among doctors themselves, NHS psychiatrists held the strongest position throughout, resting on their claims of formal specialist expertise, in spite of the lack of research-tested treatment in the early years. GPs gradually increased their status and influence in drug policy and the health service more widely, demanding recognition for their own specialist skills in the late 1990s. Nurses, psychologists, social workers and pharmacists played their role as well as representatives from the largely non-medical voluntary sector. Private prescribers, despite gaining access to policy-making circles at least in the 1980s, failed to win any concessions.

Senior civil servants have been crucial to decision-making throughout. These highly knowledgeable bureaucrats at the Department of Health and the Home Office, whether medically qualified or not, held private views on appropriate prescribing and how treatment services should be arranged which they did not usually openly declare but which guided the policies they initiated and pursued. Methods used by civil servants to push forward their aims included briefings to ministers, encouraging ministers to fund particular types of services, provision of informal advice and information to clinicians, informal support for clinicians' own political activities and, particularly important in the prescribing debates, advice and support to expert committees such as the Advisory Council on the Misuse of Drugs and good practice Guidelines

Working groups. Changes in these posts gave their opponents opportunities to make new alliances with their replacements in order to pursue alternative treatment policies.

Ministers intervened decisively at particular moments, whether to find funding for a new initiative or to veto policies developed by alliances of certain doctors and civil servants. Until the 21st century they were content to leave the details of treatment modalities to civil servants' and doctors' discretion but this hands-off approach ended in 2002 when David Blunkett, then Labour Home Secretary, called for more heroin prescribing.[1] In 2010 the Conservative-led Coalition government moved in the opposite direction to favour abstinence-based treatments over maintenance prescribing and harm reduction. However, Tory politicians may have underestimated the difficulty of determining treatment independent from the policy community and were later forced to modify their approach.[2]

Influences on these actors have been numerous. Research on the effectiveness of treatments has sometimes been used to guide decision-making but was commonly secondary to other considerations. Even in the era of 'evidence-based medicine', research findings alone have not been enough to determine which treatments were offered to patients. International developments like methadone replacement therapy were introduced from the USA, but this depended on the particular interests of certain influential doctors working in the Clinics. Wider societal changes, such as a growing scepticism about previously accepted authority and expertise, have altered the doctor–patient relationship and the balance of power between professionals and the public. The remarkable increase in the scale and spread of drug use across the country not only raised its political profile but also balanced out London's dominance over treatment and policy-making. Regional consultants were able to advance new ideas and practices, and the lack of private prescribing outside the South East of England also drew attention away from the public-private debate.

Drug treatment services have not been immune from the ongoing reorganisations of the NHS which, among other developments, strengthened primary care, formalised many treatment decisions through clinical guidelines and introduced more management. Of course the emergence of AIDS as a disease spread through shared injecting equipment and sexual contact has been of particular significance. The major shake-up of prescribing which made harm reduction a respectable guiding principle once again was not inevitably shaped by the seriousness of the virus' effects but owed a great deal to the way experts, interest groups, civil servants and government perceived the problem and interacted.[3]

The processes involving these actors and influences started out in the 1970s and 1980s with government committees, medical conferences and sometimes discussions in the medical media. From the 1980s drug use and policy were once again in the public eye with mainstream media coverage. Aside from the rhetorical debates about issues such as links between drug dependence and acquisitive crime, much of drug treatment policy continued to be made in private meetings between doctors and civil servants. The minutes and discussions of the ACMD were subject to the Official Secrets Act throughout the period. The licensing recommendations of 1984 and 1999 went directly to ministers with either no mention or only a brief reference in the published guidelines. Particularly noteworthy was the Home Office Drugs Inspectorate, barely heard of outside government except among those doctors and a few patients who had dealings with it. The Inspectorate formed its own policies without any public consultation and unlike the General Medical Council, its Tribunals were held in private.

All of these processes, influences and actors can be seen at work in the bitter conflict at the centre of *The Politics of Addiction*. Fought out in various arenas of regulation, the battle over private prescribing concerned only a handful of doctors but made ripples throughout drug treatment policy in England between 1980 and the early 21st century. In fact, a new form of medical regulation, the first official clinical guidelines, was created as a weapon in this conflict. Starting from the late 1960s, the relationship between private prescribers and NHS drug doctors went through five discernible phases. During stage one, which lasted from the late 1960s until the mid-1970s, the new Clinics were still experimenting with their approaches to treatment and private and NHS prescribers were able to co-exist fairly peacefully. Clinic psychiatrists' criticism of other doctors, if it was expressed, tended to be focused on GPs.[4] In stage two, however, as treatment preferences solidified in the mid-1970s and the Clinics became less generous in their prescribing, the contrast with private prescribers heightened and tensions arose. The Clinics saw their process of encouraging and coercing new patients into abstinence through their united detoxification front as under threat from the offer of larger scripts from doctors outside. Conversely the less appealing regime of the Clinics led more patients to seek treatment from other doctors. With the influx of large amounts of trafficked heroin at the end of the 1970s and the consequent spread of addiction, more patients than ever started to look for treatment outside the Clinics.

The third phase dates from 1983 to 1987 when regulatory battles started in earnest. The GMC launched disciplinary cases against Ann Dally and the Department of Health oversaw the development of the first clinical guidelines, along with another unsuccessful attempt to restrict prescribing of opioids for the treatment of addiction to Clinic doctors alone. Until 1987, those advocating abstinence oriented treatments were dominant in the policy community but with the emergence of HIV/AIDS, the opportunity arose for supporters of harm reduction to take the lead. From 1988 to 1999, the fourth phase, attacks on private doctors continued but maintenance prescribing was looked upon more tolerantly by regulatory authorities, the media and the public. However, the fifth and final phase from the year 2000 saw a series of disciplinary cases brought by the GMC which decimated private prescribing. Of the 11 practising private prescribers interviewed for this book between 2000 and 2003 who constituted the majority of private prescribing provision, at least six had been struck off the medical register by 2011.[5] According to one drug doctor, little private prescribing remains: 'one or two very high prescribers have appeared from time to time but been quickly quashed.'[6]

Considering the intensity of feeling evoked by private prescribers among their powerful opponents and the ineffectiveness of private doctors' organisations, it is perhaps surprising that their opponents' victory was so long in coming. The exact manoeuvres behind the GMC's assault on private prescribing in the 21st century are not known but the survival of these doctors through the 20th century can be illuminated by looking at the changing policy process. The policy community acted unwittingly as a system of checks and balances against the supremacy of any one faction, with no single interest strong enough to make policy alone. Home Office and Department of Health civil servants usually needed the support of doctors to push through their policies but these efforts could founder without ministerial support. Senior addiction psychiatrists could fend off control by civil servants but rarely achieved major policy change without their support. Civil servants usually required the support of ministers, who in turn depended on them for information and advice. When successfully allied, even if differently motivated, civil servants and consultant psychiatrists usually managed to increase regulation over other doctors during the 1980s. This happened with the production of the first clinical guidelines, which were subsequently used by both the GMC and Home Office in their disciplinary cases and in disciplinary cases against particular private doctors. However, the alliance did not give the consultants all that their leaders wanted: they had actually been seeking statutory controls over other prescribers and probably agreed to a civil servant's proposal of guidelines as a compromise.

Equally, senior civil servants in the Home Office and Department of Health were aware of a number of constraints on their own actions. Doctors' claims to clinical autonomy allowed them certain privileges in determining prescribing policies independent of other groups but by the 1990s these held less force. In the 1980s, civil servants trod a careful path to appear non-partisan in the public-private debate and were sensitive not to antagonise the powerful London consultants, who helped scupper one senior civil servant's attempts to link with private interests and build a new policy community. The Inspectorate in particular, through the 1980s and 1990s, was also sensitive to the potential conflicts of interest between regulating the doctors outside the Clinics and its role in encouraging doctors' involvement with this unpopular patient group and cultivating medical contacts for intelligence gathering.

During the 1990s, the balance between civil servants and the rest of the policy community changed. Major players at both the Department of Health and Home Office were intent on greater regulation of prescribing. Yet the voices of addiction psychiatrists had not only to compete with other professionals but also with GPs and public health doctors, all now inside the drug treatment policy community. Post-HIV, the community response to drug problems was further strengthened and hospital doctors, although still clinging to the title of 'addiction specialists', had greater difficulty controlling what went on outside. In the late 1980s and 1990s, there was greater accord in the policy community than during most of the 1980s, partly with the precarious harm-reduction consensus and also with the forced departure of vocal critic Ann Dally. These developments both helped and hindered the senior civil servants who at that time were trying to gain greater central control over prescribing. When politicians broke the cross-party consensus in the 21st century and tried to divide prescribing on party political lines, they too failed to gain total control of treatment policy.

In the early 1980s, once the London consultants had used their collective strength to move away from liberal prescribing across the board and NHS GPs were becoming re-involved in drug treatment, the consultants began to face competition and criticism from NHS GPs, private prescribers and voluntary services. These in turn gained support from particular civil servants keen to see greater diversity of provision beyond the Clinics. Although more integrated into the state than private prescribers, as contractors to the health service, GPs were nonetheless more independent than the salaried psychiatrists and shared some of the interests of each. By the late 1990s, some GPs in the policy community were acting as proxies for the excluded private prescribers, fearing that the proposed changes affecting both groups of doctors could increase centralised controls, curbing their own and their patients' autonomy.

With greater GP involvement and the diminishing impact of private practice in the policy community, the debate over appropriate prescribing moved from the language of 'public' versus 'private', and 'inside' or 'outside' the Clinics characteristic of the 1980s, to one increasingly expressed through a distinction between 'specialists' and 'generalists' in the 1990s. The specialist-generalist differentiation was used in Home Office disciplinary proceedings at this time and was a major subject of debate in the 1999 Guidelines Working Group meetings, using training and experience as the measure of specialism, rather than specialty or location of work. This partly reflected wider changes in Department of Health policy for the NHS, with its promotion of 'shared care' between specialist and primary care and a 'primary care-led' NHS. However, this essentially came down to the same fundamental issues: the control of particular prescribing practices outside the Clinics.

The policy process observed in the regulation of controlled drug prescribing, rather than demonstrating an authoritarian efficiency, is more reminiscent of Charles Lindblom's 'muddling through'; an unscientific, subjective approach where public administrators assess politically feasible policy options against previous experience and the multiple interests of pressure groups.[7] The role of civil servants in the policy community demonstrated a complex relationship between the state and the medical profession that did not fit what might be called a Foucauldian model. According to such a model, the central government inspectorate might have been expected to have taken an impersonal approach to monitoring its subjects but the changing leadership of the Inspectorate showed a very personal imprint left on the surveillance and disciplinary processes. Far from seeking conformity, Spear encouraged heterogeneity in treatment services and regulatory methods while making opportunistic alliances to achieve this. David Armstrong's Foucauldian concept of the 'Infirmary' fitted the Inspectorate's mode of working more closely than Foucault's original 'Panopticism' but does not add significantly to understanding the role of the Inspectorate.

The independent expertise that characterised the Inspectorate further developed Gerald Rhodes' findings of central government inspectorates' tendencies to diversify their roles to develop their own knowledge bases and act as advisors to ministers on policy issues. The Home Office not only enforced standards but set them too, its inspectors making a distinction between those they inspected on a friendly basis to gather intelligence who were considered reputable and those they classed as dishonest who were confronted. A Weberian bureaucracy, the Home Office Inspectorate's power was derived from technical expertise and knowledge developed through experience in the service and had a tendency to self-perpetuate into a permanent institution through adapting its aims, rather than serving the ends for which they were originally designed. These included a source of policy advice for government, training for prescribers, an occasional referral agency for patients and policy actor in its own right. Spear's behind-the-scenes involvement with AIDA and invitation to host its meetings at the Home Office conformed to a pattern found elsewhere by historian Virginia Berridge with the pressure group Action on Smoking and Health. In both cases the government was supporting voluntary organisations who could advocate policy positions desired within government but not deemed acceptable or advantageous for government to express itself.[8]

Medical self-regulation in England has been based on the idea of professional consensus, something lacking from the drugs field, particularly from the mid-1970s, with the definition of 'good practice' highly contentious both among professionals and patients. Regulation, both within the profession and by the state, was used repeatedly in attempts to control the range of drug services provided and the content of treatment with little or inconsistent use of research evidence. The GMC, the Inspectorate and the working parties which produced the clinical guidelines all formed part of the formal regulation of drug doctors representing both state-sponsored self-regulation and direct regulation by the state. Formal regulation did not use fixed rules but instead was flexible and beset by ambiguity. During the 1970s and 1980s the GMC avoided issuing specific advice, definitions, or clear rules of conduct and this was exploited by particular factions wishing to use self-regulation for their own ends.

Alongside these formal systems were informal ones which were less overt in their regulatory aims but which also combined state and self-regulation. AIDA, the AIP and the LCG were developed by the London Clinic doctors, private prescribers and NHS GPs with the encouragement of civil servants. Comparison of these three groupings using Cultural Theory has shown how and why the London consultants succeeded in fending off outside regulation and set the standards by which other doctors were judged, while the private doctors succumbed to extensive discipline. This method of social classification, developed particularly by Mary Douglas, uses the two dimensions of 'grid' (the intensity of social control to which a person is subjected) and 'group' (the degree of commitment to a social group to which someone belongs and by which his or her actions are shaped) to map social organisations and explain behaviour and beliefs.

As 'low grid, low group' individualistic entrepreneurs (a term that does not necessarily imply a profit motive), private prescribers had little social stratification and considerable individual autonomy over their own behaviour and status. Their sense of group identity and boundaries with the outside were weak, meaning that there were few controls on who could become a private prescriber or a member of AIDA or the AIP, although they did expel members for failing to conform to their practice standards. Economic competition between AIP members, with attendant issues of commercial confidentiality, led to a lack of trust and an inability to co-operate. The more hierarchical London consultants ('high grid, high group') had a strong sense of stratification and shared identity, holding in check rivalries and resentments in the overriding interest of the group. There was competition between them for prestige and resources but not for patients and they were able to share information more freely among themselves and still keep their secrets. The author was unable to access the papers of the LCG because a lone member of the LCG felt unable to act autonomously without the consent of its current membership. In this way the Group succeeded in controlling information that dated back to the 1970s, across generations, despite the fact that it was not held centrally.

Unlike the LCG, AIDA and the AIP did not try to regulate other groups of doctors but intended to raise standards among their own members and defend themselves from attack. Both AIDA and the AIP failed, collapsing from insufficient support and interest. Other than expulsion, they lacked sanctions that could be applied to non-conforming members and so could not enforce their policies. However, the LCG also lacked formal sanctions but succeeded in influencing its own members' practice and wider policy, at least during the 1970s and 1980s, so what made the difference? Patterns of involvement with the state and establishment bodies among these three groups affected their ability to influence prescribing policies and regulation. Individual personalities also played a role but secondarily to the social organisation of the associations, the economic positions of their members and their resulting values, priorities and perceived interests.

The LCG's strengths lay less in its own meetings than in members' perceived shared identity and their networks of mutual ties; bonds which integrated them both with each other and into establishment bodies inside and outside the state. Many of the LCG members also belonged to at least one of the GMC, Royal College of Psychiatrists, Advisory Council on the Misuse of Drugs and government working groups, so they encountered each other across a variety of settings. From 1977 their meetings took place at the Home Office, a location they denied to AIDA. In contrast, the private prescribers generally only saw each other at AIDA or the AIP. They rarely had other links with each other and also lacked the membership of establishment bodies, with the exception of their short time on the 1984 Guidelines Working Group. While LCG members had a lot to lose from ongoing feuds with each other, AIP and AIDA members risked less by staying true to their own preferences, if only in the short-term. Similarly, private prescribers and NHS consultants rarely met each other. What might be seen as the corporate weakness of the private doctors, they themselves would have valued as the freedom to follow their own prescribing preferences and patient needs, independent of the peer pressure that affected the London consultants. Their independence, although helping them to resist both formal and informal regulation in the short term, weakened their claims to self-regulation. Ultimately it led to a loss of freedom when formal regulation forced them to conform or stop practising. The conformity of the London consultants should not be overstated, though. Although they pushed through restricted prescribing for new Clinic patients in the mid-1970s, they retained considerable autonomy for themselves in dealing with established patients. In the 1990s, they were offering a range of different treatments, including injectable prescribing, heroin and amphetamines at their own discretion while still attempting to prevent such practices outside of the Clinics.

Their different use of the media also reflected the social characteristics of private prescribers and the London consultants. As opportunistic networkers, private prescribers saw no problem with making links outside of medicine to garner support. In the 1980s Dally drew the prescribing debate into the public sphere, a time when drug use was once more becoming prominent as a topic of national discussion. Others, such as John Marks, an NHS consultant psychiatrist famous for his advocacy of substitute heroin prescribing, and private prescriber Colin Brewer, continued to promote their views in the media in the late 1980s and 1990s and gained similar unpopularity with the Clinic establishment. Given the behind-the-scenes nature of much policy-making in drug treatment, drawing the prescribing debate outside medicine and into the public's gaze may have been seen by the London consultants as particularly reprehensible. Of course, the two cases that brought Ann Dally before the GMC provoked a considerable amount of coverage in both the medical and general media and drew scrutiny of the Council's processes, seen by several commentators as unfair. However, while Dally was able stir up controversy and also garner some sympathy for her predicament, public debate seemed to have a limited impact on the regulatory process.

Aside from one letter to *The Times* by a group of consultants and voluntary sector organisations written in 1981, Dr Dally's chief opponents, Drs Connell and Bewley, restricted their expressions to the medical press. Dally's much more public attacks on the Clinics rarely received a direct response in the non-medical public sphere. With their strong boundary against outsiders the consultants preferred to keep debate within the medical profession and in private meetings with government. The result was a more 'private' debate sought by the public doctors, who wished to keep the general public, including patients, out of the issues, and a 'public' debate pursued by private doctors.

Working to its own agendas, the general media, and particularly the tabloids, had their own impact on regulation. As well as featuring letters from both sides, a number of 'stings' were carried out by undercover reporters posing as drug-dependent patients to test the ease with which they could obtain drugs from private doctors. The resulting articles in the tabloid press prompted investigations by the Drugs Inspectorate and were also featured in disciplinary cases before the GMC. The stings continued into the 1990s and one by the *News of the World* in 1996 which encouraged a poor image of private prescribers acted as a spur to the creation of the Association of Independent Prescribers as a means of self-defence and improving their public profile.[9,10]

Considering the rise and fall of the private prescriber since 1970, there is a danger that the political nature of many of the attacks upon them and the sometimes underhand methods used to control them, could obscure any legitimate concerns about the treatment approaches of some private doctors themselves. Behind closed doors, some private prescribers expressed alarm about certain of their peers[11] and when trying to regulate themselves, expelled offending members from their associations. Throughout the 1980s and into the 1990s psychiatrists leading the London Clinics pressed policy-makers for greater control over controlled drugs at the expense of other doctors, justifying this with concerns about the influence of fee-payment on prescribing. However, this research has shown that in a private meeting with civil servants, the Clinics had first attempted to extend their prescribing powers as far back as 1969 soon after they were first set up, then citing GPs' prescribing as justification.[12] This changing target leaves an impression that the Clinics' motives were primarily imperialist in nature, weakening their later arguments that private prescribing required special controls due to the fact that they were paid directly by patients or that patients must be selling their drugs to pay fees.

The infamous disciplinary cases brought against Ann Dally also did little to help the reputation of her critics as fair or justified. Incompetence, inconsistency and a lack of clarity characterised the procedures of the GMC in the Dally cases. However, these proceedings also overshadowed the contemporaneous case of the lesser known private prescriber Peter Herman Tarnesby where real patient neglect and other poor practice came to light. On the other hand, with the self-regulation allowed to the London Clinic consultants, it is unknown whether poor practice existed in their ranks and simply failed to come to light.

At the root of much of the conflict between doctors in addiction throughout this period was the recurring tension arising from their dual role as both provider of medical care to individuals and as gatekeepers of the controlled drug supply. The only professionals in the drug treatment field able to prescribe these sought-after commodities, they found themselves paradoxically powerful and vulnerable, at risk both from exploitation by those patients only interested in securing a supply and from regulatory bodies watching the passage of drugs to patients and beyond.[13] Couched either in terms of public health, preventing the spread of addiction, or crime prevention through curbing the illegal trade in pharmaceutical drugs, doctors' policing roles became more explicit in the 1980s through the actions of the GMC. The Inspectorate had prosecuted doctors for 'irresponsible prescribing' before this but in 1983 the GMC made clear that it expected doctors to predict the likelihood of their prescribed drugs being diverted to other users, holding Dr Dally to account for the sale of drugs that she might or might not have prescribed. The regulation of doctors' prescribing was therefore also the regulation of patients' drug use. Recommendations for consumption of methadone doses to be supervised by pharmacists and picked up daily, for instance, made in the 1999 *Guidelines*, aimed both at preventing individual binge use and overdoses and stopping supplies reaching unintended hands. However, the inconvenience of daily dispensing could also interfere with other goals of recovery, such as gaining and maintaining employment. For private doctors particular issues could arise over patients not paying their fees and concerns had been expressed that payment of fees could damage the doctor–patient relationship.[14] To discharge a patient solely on grounds of non-payment would according to this logic seem unethical but the regulatory bodies' concerns over the illicit sale of prescription drugs created a situation where a doctor was at risk of disciplinary action if he or she kept treating such a patient. If a patient could not pay his or her fees, it could be argued, they would have to sell on some of their prescribed drug supply to others to afford treatment.

A remarkable feature of this story is the absence of patients from these debates. Patients were not invited onto committees such as the ACMD or the three Clinical Guidelines Working groups between 1984 and 1999. Occasionally a letter would be published in the medical press from a drug user but this was the exception.[15] Ann Dally attempted to give private patients a voice in the debate when she started up the Alba Association[16] but it failed to flourish. Other drug user groups active in the 1980s were allocated funding from a BBC appeal,[17] but also fell by the wayside. During this time the non-medical voluntary sector, who were traditionally more open to the viewpoint of drug users, and occasionally particular GPs, took on the role of 'patient advocate' within the policy community. In the 1970s and 1980s when patients were less involved in policy-making or regulation across healthcare their absence is perhaps less surprising but by the 1990s service users, including psychiatric patients, were being invited onto other health service committees. Addict patients' stigmatised status, lack of funding and poor collective organisation left them without a strong voice. Perceived by some members of the public and policy-makers as a socially deviant, criminal population that required control as much as treatment, decisions were often taken out of their hands.

Things started to change when a number of user groups emerged in the late 1990s, amid global user movements, and began to receive official recognition and support.[18,19] Email and the internet facilitated organising for those with few resources and the lessening stigma around addiction may have helped users to speak openly. The National Treatment Agency for Substance Misuse was set up by government in 2001 and appointed staff from the non-medical voluntary sector who strongly encouraged user involvement in services and planning.[20,21] At the same time, resistance to users' influence persisted, for instance, from within the Home Office's Drugs Inspectorate[22] and from some addiction psychiatrists. Eventually in 2007 even the Clinical Guidelines Working Group included user representation.

In fact, patients have probably had greatest impact on services by acting as consumers, forcing policy-makers to respond to their actions. The move out of the Clinics in the late 1970s and early 1980s was led in part by patient dissatisfaction and users sharing injecting equipment ultimately pushed services into providing needle exchanges and methadone maintenance in the wake of AIDS. While some members of the policy community have been notable for their compassion and caring, the overall impression of driving forces behind many treatment policies has been of manipulation and containment. Amid all the battles over private prescribing, the question arises of whether private patients' interests were served when their doctors were struck from the medical register or disciplined. In the 1980s, the Home Office Inspectorate had informally acted to find doctors, both private and NHS, willing to take on patients of doctors no longer able to prescribe, but once the Inspectorate ceased to deal with prescribing, the GMC did not take over responsibility towards such patients and their fate during the GMC actions in the 21st century is unknown. Heated debate about diverted pharmaceuticals from private patients led one frustrated drug doctor to comment '... it's not as if anybody really cares about these people anyway. People get really hot under the collar but what is the rest of society doing to help these people? Absolutely zero!'[23]

Beyond the world of NHS and private controlled drug prescribing, what can this interesting story tells us? Medical regulation, as seen here, was not value-neutral and parts of the process were 'captured' by particular interests. The process which has become known as 'regulatory capture', described originally by economist George Stigler, is one in which a government agency intended to regulate an industry becomes influenced by it to the extent that the agency's policies become subservient to the interests of the industry through financial corruption and other inducements.[24] While the Home Office Drugs Inspectorate was remarkably resistant to capture, aspects of the Department of Health's process, namely the clinical guidelines, were not. There was no evidence of financial inducements or corruption but the model of 'self-regulation' which underpinned these guidelines allowed particular doctors to capture the regulatory process while they themselves enjoyed freedom from its controls. This may also have been the case for the GMC but it has not been possible to determine as its decision-making processes have been hard to access.

In terms of current debates about the inclusion of private healthcare provision within the NHS, the public-private conflict in controlled drug prescribing is actually less instructive than might appear. Rather than a simple dichotomy of public versus private, the relationships between the various drug doctors tell us more about understanding organisational structures, access to particular commodities and the participation of patients in decision-making. The unresponsive nature of the Clinics in the late 1970s and 1980s reflected their hierarchical willingness to follow and impose rules, their near monopoly of treatment and their ability to keep patients out of the decision-making in services and treatment. But this does not advance the cause of private capital in the NHS; a similar situation might arise with a private hierarchical corporation monopolistically dominating treatment provision. It was rather the individualistic approach of both private prescribers *and* particular NHS GPs that made them both more resistant to regulation and more responsive to individual patients.

Appendix: Interviewed Doctors' Professional Roles

(Interviews conducted between 2000 and 2003)

Table A.1 Specialism and sources of funding by individual doctor ($n = 27$)

Private GP	Private addiction psychiatrist	NHS GP	NHS drug addiction psychiatrist	NHS non-addiction psychiatrist
✓			✓	
✓				
✓		✓ (previously)		
✓				
✓		✓		
		✓		
	✓		✓ (previously)	✓ (previously)
	✓		✓ (alcohol addiction psychiatry, previously)	
		✓		
	✓			
✓		✓		
	✓		✓	
	✓			✓
✓				
			✓	
		✓		
	✓		✓	
			✓	
		✓		
✓				
			✓	
			✓	
		✓		
			✓	
			✓	
			✓	
			✓	

Table A.2 Number of doctors funded from different sources ($n = 27$)

Roles of doctors interviewed	Working at time of interview	Retired at time of interview	Totals
Solely Private GPs	3	2	5
Solely NHS Addiction Psychiatrists	4	4	8
Solely NHS GPs	5	0	5
Solely Private Addiction Psychiatrists	1	0	1
NHS and Private GPs	2	0	2
NHS and Private Addiction Psychiatrists (2 of 4 had left the NHS before practising privately)	4	0	4
Private addiction psychiatrists and NHS non-addiction psychiatrists	2	0	2

Notes

Introduction

1. V. Berridge, *Opium and the People. Opiate Use and Drug Control Policy in Nineteenth and Early Twentieth Century England* (first published 1981, London: Allen Lane; this edition London: Free Association Books, 1999).
2. Departmental Committee on Morphine and Heroin Addiction, Report [Rolleston Report] (London: HMSO, 1926), pp. 6–7.
3. D. Musto, *The American Disease. Origins of Narcotic Control* (New York and Oxford: Oxford University Press, 1999, third edition, first edition 1973), pp. 121–150.
4. M. Sharpe, Interview by Sarah Mars (2001).
5. R. Lewis, 'Flexible hierarchies and dynamic disorder – the trading and distribution of illicit heroin in Britain and Europe, 1970–90', in J. Strang and M. Gossop (eds.), *Heroin Addiction and Drug Policy: The British System* (Oxford, New York, Tokyo: Oxford University Press, 1994), pp. 42–54.
6. See J. Merrill, 'Dexamphetamine substitution as a treatment for amphetamine dependence', The First National Conference on Stimulants, Manchester Metropolitan University, 18 September 1998.
7. S. Morris, 'Doctors accused on heroin advice', *The Guardian* (24 February 2004).
8. For example, Advisory Council on the Misuse of Drugs, *Treatment and Rehabilitation*, DHSS (London: HMSO, 1982); Anonymous, 'Doctor Death', *The Listener* (29th July 1982), 22.
9. For example, A. Dally, *A Doctor's Story* (London: Macmillan, 1990), pp. 57–98.
10. T. Bewley, and A. H. Ghodse, 'Unacceptable face of private practice: prescription of controlled drugs to addicts', *British Medical Journal*, 286, (1983), 1876–1877.
11. 'File on Four', BBC Radio Four (1997).

1 1965–2010: A Background Sketch

1. The main sources of information about the availability of drugs and the number of drug users in treatment were those compiled by the Home Office. Drug availability was gauged through the number of seizures of drugs both at borders and within England by enforcement agencies. As a measure of drugs available in England this was far from accurate. Shortcomings of the data and caveats for interpretation have been described elsewhere – for example, Institute for the Study of Drug Dependence, *Drug Misuse in Britain 1996* (London: ISDD, 1997). However, as an indicator of relative increases, it has proved valuable.

2. Between 1968 and 1997 it was a statutory requirement for doctors treating patients dependent on opiates or cocaine to notify the Home Office's Addicts Index. Although methods of data collection changed over this period and may not have been comprehensive or entirely accurate (for reasons such as the use of false names by drug users or doctors' failure to notify the Index), this was considered the best source for comparisons over more than a decade and gave an indication of the vast increase in the number of addicted patients (see J. Mott, 'Notification and the Home Office', in J. Strang and M. Gossop (eds.), *Heroin Addiction and Drug Policy: The British System* (Oxford, New York, Tokyo: Oxford University Press, 1994), pp. 270–291).

3. J. Mott (1994).

4. V. Berridge, *AIDS in the UK. The Making of Policy, 1981–1994* (Oxford: Oxford University Press, 1996), p. 91.

5. *Ibid.* pp. 119–121, 221.

6. H. Hudebine, 'Applying cognitive policy analysis to the drug issue: harm reduction and the reversal of the deviantization of drug users in Britain 1985–2000', *Addiction Research and Theory*, 13 (2005), 231–243.

7. H. B. Spear (and ed. J. Mott), *Heroin Addiction Care and Control: The 'British System' 1916–1984* (London: Drugscope, 2002).

8. C. Smart, 'Social policy and drug dependence: an historical case study', *Drug and Alcohol Dependence*, 16 (1985), 169–180.

9. V. Berridge, *Opium and the People* (first published 1981 London: Allen Lane; this edition London: Free Association Books, 1999).

10. J. Strang, ' "The British System": past, present and future', *International Review of Psychiatry*, 1 (1989), 109–120.

11. G. V. Stimson and R. Lart, 'The relationship between the state and local practice in the development of national policy on drugs between 1920 and 1990' in J. Strang and M. Gossop (1994), pp. 331–341.

12. Interdepartmental Committee on Drug Addiction, *Report of the Interdepartmental Committee on Drug Addiction* (London: HMSO, 1961), quoted in Interdepartmental Committee on Drug Addiction, *Drug Addiction. The Second Report of the Interdepartmental Committee* [second Brain Report], Ministry of Health, Scottish Home and Health Department (London: HMSO, 1965), p. 4.

13. Interdepartmental Committee on Drug Addiction (1965), p. 8.

14. D. Hawks, 'The dimensions of drug dependence in the United Kingdom', in G. Edwards, M. A. H. Russell, D. Hawks *et al.* (eds.), *Drugs and Drug Dependence* (Farnborough, Hants., England: Saxon House and Lexington, MA: Lexington Books, 1976), p. 7.

15. *Ibid.* p. 7.

16. Interdepartmental Committee on Drug Addiction (1965), p. 5.

17. *Ibid.* p. 6.

18. G. V. Stimson and E. Oppenheimer, *Heroin Addiction: Treatment and Control in Britain* (London: Tavistock, 1982), pp. 49–53.

19. C. Smart, 'Social policy and drug dependence: an historical case study', *Drug and Alcohol Dependence*, 16 (1985), 169–180.

20. A. Glanz, 'The fall and rise of the general practitioner', in J. Strang and M. Gossop (1994), pp. 151–166.

21. T. H. Bewley and R. S. Fleminger, 'Staff/patient problems in drug dependence treatment clinics', *Journal of Psychosomatic Research*, 14 (1970), 303–306.
22. H. B. Spear (and ed. J. Mott) (2002), p. 170.
23. They had been included in legislation between 1926 to 1961 but had not been used – see P. Bean, 'Policing the medical profession: the use of tribunals' in D. K. Whynes and P. T. Bean (eds.), *Policing and Prescribing. The British System of Drug Control* (Basingstoke: Macmillan, 1991) pp. 60–70. – and are discussed in detail in Chapter 5.
24. D. Hawks, 'The dimensions of drug dependence in the United Kingdom', in G. Edwards, M. A. H. Russell, D. Hawks *et al.* (eds.), *Drugs and Drug Dependence* (Farnborough, Hants., England: Saxon House and Lexington, MA: Lexington Books, 1976), pp. 5–29.
25. A. Glanz (1994), p. 155.
26. H. B. Spear (and ed. J. Mott) (2002), p. 228.
27. R. Lewis, R. Hartnoll, S. Bryer *et al.*, 'Scoring smack: the illicit heroin market in London 1980–1983', *British Journal of Addiction*, 80 (1985), 281–290.
28. D. Turner, 'The development of the voluntary sector: no further need for pioneers?', in J. Strang and M. Gossop (1994), pp. 222–230.
29. Under the law, doctors were obliged to notify to the Home Office Addicts Index anyone they attended who was dependent on certain specified opiates or cocaine. Doctors were encouraged to phone the Addicts Index to find out if anyone for whom they were about to prescribe a drug was already receiving a prescription from another doctor.
30. ACMD, *Treatment and Rehabilitation*, DHSS (London: HMSO, 1982), p. 120.
31. B. Wells, 'Narcotics Anonymous (NA) in Britain', in J. Strang and M. Gossop (1994), pp. 240–247.
32. J. Strang (1989), p. 113.
33. DHSS, *Better Services for the Mentally Ill* (London: HMSO, 1975) pp. 68–69.
34. C. Webster, *The National Health Service. A Political History* (Oxford: Oxford University Press, 1998), p. 138.
35. *Ibid.* pp. 138–139.
36. *Ibid.* p. 111.
37. Anonymous, 'What's happening with heroin?' *Druglink Information Letter*, ISDD, 17 (1982), 1–5.
38. Interdepartmental Committee on Drug Addiction (1965), p. 5.
39. M. Ashton, 'Controlling addiction: the role of the Clinics', *Druglink Information Letter*, ISDD, 13 (1980), 1–6.
40. For example, P. H. Connell, '1985 Dent Lecture: "I need heroin". Thirty years' experience of drug dependence and of the medical challenges at local, national, international and political level. What next?', *British Journal of Addiction*, 81 (1986), 461–472.
41. V. Berridge (1996), p. 92.
42. A. Glanz (1994), pp. 155–156.
43. ACMD (1982), pp. 81–86.
44. N. McKeganey, 'Shadowland: general practitioners and the treatment of opiate-abusing patients', *British Journal of Addiction* 83 (1988), 373–386.
45. P. M. Strong, 'Doctors and dirty work – the case of alcoholism', *Sociology of Health and Illness*, 2 (1980), 24–47.

46. H. Ghodse, A. Oyefeso and B. Kilpatrick, 'Mortality of drug addicts in the UK, 1967–1993', *International Journal of Epidemiology*, 27 (1998), 473–478.

47. A. H. Ghodse, 'Drug problems dealt with by 62 London casualty departments', *British Journal of Preventive and Social Medicine*, 30, 4 (1976), 251–256.

48. A. Glanz (1994), p. 155.

49. G. V. Stimson and E. Oppenheimer (1982), p. 49.

50. H. B. Spear (2002), p. 258.

51. Working Party of the Royal College of Psychiatrists and the Royal College of Physicians, *Drugs, Dilemmas and Choices* (London: Gaskell, 2000) p. 50.

52. *Ibid.* p. 50.

53. R. Power, 'Drug trends since 1968', in J. Strang and M. Gossop (1994), pp. 29–41, 33.

54. British Medical Association, *The Misuse of Drugs* (Amsterdam: Harwood Academic Publishers, 1997) p. 22.

55. ACMD (1982), pp. 81–86.

56. S. MacGregor, B. Ettorre, R. Coomber, *et al.*, *Drug Services in England and the Impact of the Central Funding Initiative*, ISDD Research Monograph One (London: ISDD, 1991).

57. V. Berridge (1996), p. 94.

58. G. V. Stimson, 'British drug policies in the 1980s: a preliminary analysis and suggestions for research', *British Journal of Addiction* 82, 5 (1987), 477–488.

59. D. Turner (1994), p. 222.

60. A. Glanz (1994), pp. 155–158.

61. J. Strang, 'A model service: turning the generalist on to drugs' in S. MacGregor (ed.), *Drugs and British Society. Responses to a Social Problem in the 1980s* (London and New York: Routledge, 1989), pp. 143–169.

62. For example, N. Dorn and N. South (eds.), *A Land Fit for Heroin? Drug Policies, Prevention and Practice* (Basingstoke: Macmillan, 1987).

63. S. MacGregor (ed.), *Drugs and British Society. Responses to a Social Problem in the 1980s* (London and New York: Routledge, 1989).

64. For example, V. Berridge, 'Historical issues' in S. MacGregor (1989), pp. 20–35.

65. K. Duke, *Drugs, Prisons and Policy-Making* (London: Macmillan, 2003).

66. G. V. Stimson (2000), pp. 259–264.

67. See A. Glanz (1994), p. 162.

68. *Ibid.* p. 162.

69. British Medical Association. General Medical Services Committee, *Core Services: Taking The Initiative* (London: British Medical Association, 1996).

70. A. Glanz (1994), p. 159.

71. R. Klein, *The New Politics of the NHS* (first published 1983, London and New York: Longman; fourth edition, Harlow: Prentice Hall, 2001).

72. V. Berridge (1996), pp. 93–94.

73. *Ibid.* pp. 220–225.

74. V. Berridge, 'AIDS and British drug policy: continuity or change?' in V. Berridge and P. Strong (eds.), *AIDS and Contemporary History* (Cambridge: Cambridge University Press, 1993), pp. 135–156.

75. D. Turner (1994), pp. 225–226.

76. V. Berridge (1993), pp. 146–148.

77. ACMD, *AIDS and Drug Misuse* (London: DHSS, 1988).
78. For example, A. Preston and G. Bennett, 'The history of methadone and methadone prescribing', in G. Tober and J. Strang (eds.), *Methadone Matters: Evolving Community Methadone Treatment of Opiate Addiction* (London: Martin Dunitz, 2003), pp. 13–20, 19.
79. V. Berridge (1996), p. 92.
80. V. Berridge (1993), p. 146.
81. H. Hudebine.
82. Department of Health, Scottish Office Home and Health Department, and Welsh Office, *Drug Misuse and Dependence, Guidelines on Clinical Management* (London: HMSO, 1991).
83. H. Hudebine.
84. J. Strang, G. V. Stimson and D. C. Des Jarlais, 'What is AIDS doing to the drug research agenda?', *British Journal of Addiction*, 87 (1992), 343–346.
85. V. Berridge (1996), p. 276.
86. H. Hudebine.
87. *Ibid.*
88. N. South, 'Tackling drug control in Britain: from Sir Malcolm Delevingne to the new drugs strategy', in R. Coomber (ed.), *The Control of Drugs and Drug Users. Reason or Reaction?* (Amsterdam: Harwood Academic Publishers, 1998), pp. 87–106.
89. An exception might be, for instance, possession of a Schedule 2, such as heroin, with a prescription.
90. For example, Labour Party, *Drugs: The Need for Action* (London: Labour Party, 1994).
91. For example, N. Dorn, O. Baker & T. Seddon, *Paying For Heroin: Estimating The Financial Cost Of Acquisitive Crime Committed by Dependent Heroin Users in England and Wales* (London: ISDD, 1994).
92. H. Parker, C. Bury, and R. Egginton, *New Heroin Outbreaks Amongst Young People in England and Wales*, Crime Detection and Prevention Series, Paper 92 (London: Home Office Police Policy Directorate, 1998), pp. 6–7.
93. G. V. Stimson (2000), pp. 259–264.
94. P. J. Turnbull, T. McSweeney, R. Webster, *et al.*, *Drug Treatment and Testing Orders: Final Evaluation Report*, Home Office Research Study 212 (London: Home Office Research, Development and Statistics Directorate, 2000), pp. 1–7.
95. H. Hudebine.
96. In 2000 Gerry Stimson helped to establish and became Chair of the UK Harm Reduction Alliance to campaign for harm reduction policies.
97. V. Berridge (1993), pp. 138–152.
98. *Ibid.* p. 152.
99. V. Berridge (1996), pp. 91–92.
100. V. Berridge, 'Issue network versus producer network? ASH, the Tobacco Products Research Trust and UK smoking policy' in V. Berridge (ed.), *Making Health Policy: Networks in Research and Policy After 1945* (Amsterdam: Rodopi, 2005), pp. 101–124.
101. D. Turner (1994), p. 228.
102. *Ibid.* p. 229.
103. B. Wells (1994), pp. 241–246.

104. For example, UK Health Departments, *Drug Misuse and Dependence: Guidelines on Clinical Management* (London: The Stationery Office, 1999), pp. 10–15.
105. G. V. Stimson and R. Lart (1994), pp. 334–335.
106. C. Webster (1998), pp. 140–214.
107. R. Klein (2001), pp. 169–172.
108. G. V. Stimson and R. Lart (1994), pp. 336–339.
109. R. Klein (2001), pp. 180–181.
110. *Ibid.* p. 210.
111. Staff and Agencies, 'Harold Shipman: a chronology', *The Guardian* (July 15, 2004), www.guardian.co.uk/archive.
112. Lord Privy Council, *Tackling Drug Misuse: A Summary of the Government's Strategy* (London: HMSO, 1985).
113. G. V. Stimson (1987), pp. 483–484.
114. Lord President of the Council and Leader of the House of Commons, Secretary of State for the Home Department, Secretary of State, *et al.*, *Tackling Drugs Together. A Strategy for England 1995–1998* (London: HMSO, 1995).
115. G. V. Stimson (2000), p. 260.
116. For example, Working Party of the Royal College of Psychiatrists and the Royal College of Physicians (2000), p. 258.
117. N. South, 'Tackling drug control in Britain: from Sir Malcolm Delevingne to the new drugs strategy' in R. Coomber (ed.), *The Control of Drugs and Drug Users. Reason or Reaction?* (Amsterdam: Harwood Academic Publishers, 1998), pp. 87–106.
118. S. MacGregor, 'Pragmatism or principle? continuity and change in the British approach to treatment and control', in R. Coomber (1998), pp. 131–154.
119. For example, G. V. Stimson (2000), pp. 259–264.
120. R. Yates. 'A brief history of British drug policy, 1950–2001', *Drugs: Education, Prevention and Policy*, 9 (2002), 113–124.
121. G. Pearson, 'Drugs at the end of the century', *British Journal of Criminology*, 39 (1999), pp. 477–487.
122. For example, G. V. Stimson and R. Lart (1994), pp. 331–341.
123. For example, V. Berridge (1993), pp. 135–156.
124. For example, C. Smart, 'Social policy and drug addiction: a critical study of policy development', *British Journal of Addiction*, 79 (1984), 31–39.
125. For example, K. Duke (2003).
126. For example, S. MacGregor (1998), pp. 131–154.
127. V. Berridge (1996).
128. For example, V. Berridge (1993), pp. 139–143.
129. G. V. Stimson (1987), p. 481.
130. *Ibid.* p. 484.
131. R. Power, 'Drugs and the media: prevention campaigns and television', S. MacGregor (1989), pp. 129–142.
132. *Ibid.* pp. 133–134.
133. S. MacGregor (1998).
134. H. Hudebine.
135. J. Hoare and D. Moon (eds.), *Drug Misuse Declared: Findings from the 2009/10 British Crime Survey England and Wales* (London: Home Office, 2010).

136. S. MacGregor, 'Policy Responses to the Drugs Problem', in S. MacGregor (ed.), *Responding to Drug Misuse. Research and Policy Priorities in Health and Social Care* (London and New York: Routledge, 2010), pp. 1–13.

2 Prescribing and Proscribing: The *Treatment and Rehabilitation* Report

1. T. Bewley, Interview by Sarah Mars (2001)
2. Department of Health and Social Security, 'Heroin Dependence: Clinical Conference' [Minutes of meeting] (18th September 1969), Private archive.
3. G. V. Stimson and E. Oppenheimer (1982), p. 99.
4. M. Sharpe, Interview by Sarah Mars (2001).
5. Department of Health and Social Security, 'Heroin dependence: Clinical conference, 1st December 1969' [Minutes of meeting] (January 1970), Private archive.
6. See also T. Bewley, 'The Illicit Drug Scene', *British Medical Journal*, 2 (1975), 318–320.
7. H. B. Spear (and ed. J. Mott), *Heroin Addiction Care and Control: The 'British System' 1916–1984* (London: DrugScope, 2002), p. 194.
8. P. H. Connell, 'Drug dependence in Great Britain: a challenge to the practice of medicine', in H. Steinberg (ed.), *Scientific Basis of Drug Dependence*, Coordinating Committee for Symposia on Drug Action (London: J&A Churchill, 1969), pp. 291–299, p. 293.
9. DHSS, *Better Services for the Mentally Ill* (London: HMSO, 1975), p. 67.
10. M. Mitcheson, 'Drug Clinics in the 1970s', in J. Strang and M. Gossop (eds.), *Heroin Addiction And Drug Policy: The British System* (Oxford, New York, Tokyo: Oxford University Press, 1994), pp. 179–191.
11. H. B. Spear (and ed. J. Mott), *Heroin Addiction Care and Control: The 'British System' 1916–1984* (London: Drugscope, 2002), pp. 275–276.
12. Controlled drugs are those controlled by the Misuse of Drugs Act, 1971.
13. T. H. Bewley, 'Prescribing Psychoactive Drugs to Addicts', *British Medical Journal*, 281 (1980), 497–498, p. 497.
14. *Ibid.*
15. Since 1968 doctors had needed a special Home Office licence in order to prescribe heroin or cocaine for the treatment of addiction. These licences had been granted almost exclusively to Clinic doctors.
16. *The Lancet*, 'Drug addiction: British System failing', *The Lancet*, 1 (1982), 83–84, p. 83.
17. T. H. Bewley (1980), p. 497.
18. For example, A. Dally, 'Personal View', *British Medical Journal*, 283 (1981), 857.
19. ACMD, *Treatment and Rehabilitation*, DHSS (London: HMSO, 1982).
20. The ACMD is an independent body set up to advise government under the Misuse of Drugs Act, 1971.
21. Senior Civil Servant, DHSS, Interview by Sarah Mars (2001).
22. ACMD (1982), pp. 46–47.
23. DHSS, *Better Services for the Mentally Ill* (London: HMSO, 1975), pp. 69–70.

24. G. V. Stimson and R. Lart, 'The Relationship Between the State and Local Practice in the Development of National Policy on Drugs between 1920 and 1990', in J. Strang and M. Gossop (eds.), *Heroin Addiction And Drug Policy: The British System* (Oxford, New York, Tokyo: Oxford University Press, 1994), pp. 331–341, p. 336.

25. see ACMD (1982), pp. 20–22.

26. For further discussion of why the Home Office rather than the DHSS provided the secretariat, see S. G. Mars (2005), pp. 63–64.

27. ACMD, Treatment and Rehabilitation Working Group, *First Interim Report* (London: DHSS, 1977).

28. For a full discussion of the first working group and its interim report, see S. G. Mars, *Prescribing and Proscribing: The Public-Private Relationship in the Treatment of Drug Addiction in England, 1970–99* (University of London: PhD Thesis, 2005), pp. 63–69.

29. *Ibid.* p. 8.

30. D. Turner [SCODA], TRWG (2)/20 'Treatment and Rehabilitation Working Group' (16th November 1978), File D/A242/12, DH Archive, Nelson, Lancashire.

31. D. Turner [SCODA] (16th November 1978).

32. *Ibid.*

33. D. Turner [SCODA], TRWG (2)/20 'Treatment and Rehabilitation Working Group' (16th November 1978), File D/A242/12, DH Archive, Nelson, Lancashire.

34. D. Turner [SCODA], Personal communication (2003).

35. That is, people visiting doctors for treatment and diagnosed as dependent on opiates or cocaine.

36. ACMD (1982), p. 45.

37. T. Bewley, Interview by Sarah Mars (2001).

38. *Ibid.* pp. 51–62.

39. *Ibid.* p. 52.

40. *Ibid.* pp. 52–53.

41. *Ibid.* p. 54.

42. *Ibid.*

43. Dipipanone combined with the anti-nausea drug cyclizine was marketed as *Diconal* and had been widely illicitly used in the north of England.

44. A. Thorley, Interview by Sarah Mars (2002).

45. H. B. Spear (and ed. J. Mott) (2002), *op cit.*, p. 63.

46. S. MacGregor, 'Choices for policy and practice', in S MacGregor (ed.), *Drugs and British Society. Responses to a Social Problem in the 1980s* (London and New York: Routledge, 1989), pp. 170–200.

47. Interdepartmental Committee on Drug Addiction, *Drug Addiction. The Second Report of the Interdepartmental Committee* [second Brain Report], Ministry of Health, Scottish Home and Health Department (London: HMSO, 1965), p. 9.

48. DHSS, *Better Services for the Mentally Ill* (London: HMSO, 1975).

49. ACMD, Treatment and Rehabilitation Working Group (1977), p. 9.

50. ACMD (1982), p. 34.

51. M. Plant, 'The epidemiology of illicit drug-use', in S. MacGregor (ed.), *Drugs and British Society. Responses to a Social Problem in the 1980s* (London and New York: Routledge, 1989), pp. 52–63.

52. Home Office, TRWG Mins 23, Minutes of the 23rd Meeting of the Treatment and Rehabilitation Working Group (27th November 1978), File D2/A242/12 Vol G., DH Archive, Nelson, Lancashire.

53. A. Thorley, Interview by Sarah Mars (2002).

54. J. Strang, 'A model service: turning the generalist on to drugs', in S. MacGregor (ed.) *Drugs and British Society. Responses to a Social Problem in the 1980s* (London and New York: Routledge, 1989), pp. 143–169.

55. B. Thom, *Dealing With Drink. Alcohol and Social Policy from Treatment to Management* (London and New York: Free Association Books, 1999), pp. 105–134.

56. ACMD (1982), p. 31.

57. Home Office, TRWG Mins 23, Minutes of the 23rd Meeting of the Treatment and Rehabilitation Working Group (27th November 1978), File D2/A242/12 Vol G., DH Archive, Nelson, Lancashire.

58. ACMD (1982), p. 35.

59. J. Strang, '"The British System": past, present and future'. *International Review of Psychiatry*, 1 (1989), 109–120.

60. V. Berridge, 'Historical issues', in S. MacGregor (ed.), *Drugs and British Society. Responses to a Social Problem in the 1980s* (London and New York: Routledge, 1989), pp. 20–35.

61. H. B. Spear (and ed. J. Mott) (2002), p. 276.

62. *Ibid.* p. 276.

63. A consultant is the most senior grade for doctors working in British hospitals. Doctors heading up the specialist Clinics would usually be consultants.

64. Home Office, TWG Mins 22. ACMD, 'Treatment and Rehabilitation Working Group', Meeting held on 2nd October 1978, File D/A242/12, DH Archive, Nelson, Lancashire.

65. D. Turner [SCODA] (2003).

66. Anonymous, 'Journal Interview 27: Conversation with Philip Connell', *British Journal of Addiction*, 85 (1990), 13–23, p. 13.

67. T. Bewley, Interview by Sarah Mars (2001).

68. A. Kerr, 'In conversation with Thomas Bewley', *Psychiatric Bulletin*, 31 (2007), 220–223.

69. R. Cawley, Obituary: Dr P H Connell, *The Independent* (Monday 10th August 1998).

70. J. Willis, Interview by Sarah Mars (2003).

71. GMC, *Minutes of the General Medical Council and Committees for the Year 1979 with Reports of the Committees, etc.* CXVI (London: GMC, 1979).

72. GMC, *Minutes of the General Medical Council and Committees for the Year 1991 with Reports of the Committees, etc.* CXXVIII (London: GMC, 1991).

73. T. Bewley, 'Prescribing psychoactive drugs to addicts', *British Medical Journal*, 281 (1980), 497–498.

74. T. Bewley (2001).

75. A. Dally, *A Doctor's Story* (London: Macmillan, 1990), pp. 141–144.

76. Home Office, TRWG (82) 24 (7th May 1982). File DAC 7, DH Archive, Nelson, Lancashire.

77. J. L. Reed, 'Meeting of Doctors Working in London Drug Dependency Treatment Centres, November 25th, 1969 at St Bartholomew's Hospital' [Minutes of meeting], Private Archive.

78. A. Thorley, TRWG (82) 15, Memorandum to David Hardwick (22nd March, 1982) File DAC 28, DH Archive, Nelson, Lancashire.
79. D. Turner (2002).
80. A. Thorley (2002).
81. ACMD (82) 2nd meeting minutes (Meeting held on 13 July 1982). File MDS/1/3 Vol 3, DH Archive, Nelson, Lancashire.
82. Home Office, TRWG Mins 23, Minutes of the 23rd Meeting of the Treatment and Rehabilitation Working Group (27th November 1978), File D2/A242/12 Vol G., DH Archive, Nelson, Lancashire.
83. A. Thorley (2002).
84. Senior Civil Servant, DHSS, Interview by Sarah Mars (2001).
85. D. Turner [SCODA] (2002).
86. *Ibid.*
87. AIDA, 'Drug Addiction: Guidelines and Standards in Management', Pre-publication edition., February 1982, File PP/DAL/B/4/1/1/5, Wellcome Library.
88. AIDA, TRWG(82)10, 'Drug Addiction: Guidelines and Standards in Management. Pre-publication edition (February 1982), File DAC 7, DH Archive, Nelson, Lancashire.
89. T. H. Bewley (1980), pp. 497–498.
90. D. Turner [SCODA] (2002).
91. ACMD (1982), p. 57.
92. *Ibid.* p. 33.
93. *Ibid.* p. 22.
94. *Ibid.* p. 57.
95. D. Turner [SCODA] (2003).
96. ACMD (82) Second meeting minutes. (Meeting held on July 13 1982). File MDS/1/3 Vol 3, DH Archive, Nelson, Lancashire.
97. A. Thorley (2002).
98. D. Turner (2002).
99. A. Thorley (2002).
100. D. Turner [SCODA], Interview by Sarah Mars (2002).
101. A. Thorley (2002).
102. *Ibid.*
103. C. Webster, *The National Health Service. A Political History* (Oxford: Oxford University Press, 1998), pp. 140–219.
104. A. M. Blythe, Memo from to J. M. Rogers (19th February 1982), File 16/DAC 28 Vol 2, DH Archive, Nelson, Lancashire.
105. ACMD: Report on Treatment and Rehabilitation. Draft Submission (November 1983) From File DAC 14 Volume 4, DH Archive, Nelson, Lancashire.
106. K. Clarke, Memo to Secretary of State (31st October 1983), File DAC 14 Volume 4, DH Archive, Nelson, Lancashire.
107. ACMD. *Treatment and Rehabilitation* DHSS (London: HMSO, 1982).
108. A. Thorley, 'Longitudinal Studies of Drug Dependence', in G. Edwards and C. Busch (eds.), *Drug Problems in Britain: A Review of Ten Years* (London: Academic Press, 1981), pp. 117–169.
109. For example, ACMD, *AIDS and Drug Misuse Part 2* (London: HMSO, 1989).
110. Senior Civil Servant, DHSS, Interview by Sarah Mars (2001).
111. A. M. Blythe, Memo to M. Moodie. 'Services for Drug Misusers. ACMD' (5th February 1982), File DAC 7, DH Archive, Nelson, Lancashire.

112. Senior Civil Servant, DHSS (2001).
113. Anonymous, 'Note of the One-day medical conference convened at the DHSS to discuss the medical response to the ACMD report on Treatment and Rehabilitation on 28th January 1983'. File DAC 28, DH Archive, Nelson, Lancashire.
114. *Ibid.*
115. L. Webb. Letter to E. Shore, Deputy Chief Medical Officer, DHSS. (12th October, 1984) File DAC 28, DH Archive, Nelson, Lancashire.
116. Anonymous, 'Note conference on 28th January 1983'.
117. *Ibid.*
118. Anonymous, Draft reply to Home Secretary from Norman Fowler (undated) File DAC 14 Vol 4, DH Archive, Nelson, Lancashire.
119. N. Fowler, Letter to L. Britton (30th November 1983), File DAC 14, DH Archive, Nelson, Lancashire.

3   Defining 'Good Clinical Practice'

1. A. Thorley, Interview by Sarah Mars (2002).
2. For example, Dally, A. 'Drug clinics today' [letter], *The Lancet*, 8328 (1983), 826.
3. *Medical Working Group on Drug Dependence, Guidelines of Good Clinical Practice in the Treatment of Drug Misuse* (London: DHSS, 1984).
4. S. Harrison and W. I. U. Ahmad, 'Medical autonomy and the UK state 1975–2025', *Sociology*, 34 (2000), 129–146.
5. A. Thorley, 'Longitudinal studies of drug dependence' in G. Edwards and C. Busch (eds.) *Drug Problems in Britain: A Review of Ten Years* (London: Academic Press, 1981), pp. 117–169.
6. M. Mitcheson, 'Drug clinics in the 1970s' in J. Strang and M. Gossop (eds.), *Heroin Addiction and Drug Policy: The British System* (Oxford, New York and Tokyo: Oxford University Press, 1994), pp. 179–191.
7. R. Hartnoll, M. C. Mitcheson, A. Battersby, G Brown, M Ellis, P. Fleming, and N. Hedley, 'Evaluation of heroin maintenance in controlled trial', *Archives of General Psychiatry*, 37 (1980), 877–884.
8. *Ibid.* p. 883.
9. *Ibid.*
10. G. V. Stimson and E. Oppenheimer, *Heroin Addiction: Treatment and Control in Britain* (London: Tavistock Publications, 1982), pp. 215–219.
11. M. Sharpe, Interview by Sarah Mars (2001).
12. T. Bewley, Interview by Sarah Mars (2001).
13. H. B. Spear (and ed. J. Mott), *Heroin Addiction Care and Control: the 'British System' 1916–1984* (London: Drugscope, 2002), pp. 245–246.
14. J. Strang, " 'The British System": past, present and future', *International Review of Psychiatry*, 1 (1989), p. 113.
15. DHSS, 'Drug dependence: clinical conference' [Minutes of meeting] (28th November 1968), Private archive.
16. H. B. Spear (and ed. J. Mott) (2002), p. 243.
17. M. Mitcheson (1994), pp. 178–179.
18. K. Sathananthan, Interview by Sarah Mars (2001).
19. J. H. Willis, 'Unacceptable face of private practice: prescription of controlled drugs to addicts' [letter], *British Medical Journal*, 287 (1983), 500.

20. Department of Health and Social Security, Department of Education and Science, Home Office and Manpower Services Commission, *Misuse of Drugs with Special Reference to the Treatment and Rehabilitation of Misusers of Hard Drugs: Government Response to the Fourth Report from the Social Services Committee Session 1984–85* (London: HMSO, 1985), pp. 18–19.
21. A. Macfarlane, Interview by Sarah Mars (2002).
22. See ACMD, *Treatment and Rehabilitation*, DHSS (London: HMSO, 1982), pp. 26–27.
23. J. Strang, 'Personal view', *British Medical Journal*, 283 (1981), 376.
24. AIDA, 'Drug Addiction: Guidelines and Standards in Management', Pre-publication edition, February 1982, File PP/DAL/B/4/1/1/5, Wellcome Library.
25. AIDA, 'Minutes of first meeting of the Working Party' (24th November, 1981), File PP/DAL/B/4/1/1/4, Wellcome Library, London.
26. W. Laing, *Going Private. Independent Health Care in London* (London: King's Fund, 1992).
27. C. Webster, *The National Health Service. A Political History* (Oxford: Oxford University Press, 1998), p. 146.
28. H. B. Spear (and ed. J. Mott) (2002), p. 37.
29. *Ibid*. p. 260.
30. T. Bewley, Interview by Sarah Mars (2001).
31. A. Dally (1990), pp. 71–275.
32. D. Turner [SCODA], Interview by Sarah Mars (2002).
33. V. Berridge, 'Doctors and the state: the changing role of medical expertise in policy-making', *Contemporary British History*, 11 (1997), 66–85.
34. For example, H. B. Spear (and ed. J. Mott) (2002), pp. 62–63.
35. For example, *Ibid*. p. 62.
36. T. Bewley (2001).
37. Home Office, Personal communication (2002).
38. Senior Civil Servant, DHSS (2001).
39. H. D. Beckett, Interview by Sarah Mars (2001).
40. D. Turner (2002).
41. For example, T. Waller and A. Banks, 'Drug abuse pull out supplement', *GP* (25th March 1983), 27–45 and A. Banks and T. A. N. Waller, *Drug Addiction and Polydrug Use: The Role of the General Practitioner* (London: ISDD, 1983).
42. For example, E. Tylden, 'Care of the pregnant drug addict', *MIMS*, 1st June (1983), 1–4.
43. A. Thorley, Interview by Sarah Mars (2002).
44. Senior Civil Servant, DHSS (2001).
45. T. Bewley and A. H. Ghodse (1983), pp. 1876–1877.
46. A. Dally (1983), p. 826.
47. Anonymous, 'Journal Interview 36: Conversation with Thomas Bewley', *Addiction*, 90 (1995), 883–892.
48. T. Bewley, 'Drug dependence in the USA', *Bulletin on Narcotics*, XXI, 2 (1969), 13–29.
49. See T. H. Bewley (1980), p. 497.
50. See T. Bewley and A. H. Ghodse (1983), pp. 1876–1877.
51. A. Banks, Interview by Sarah Mars (2001).

52. A. Banks, Letter to N. Fowler, MP (1983), Ref 40117, DrugScope Library, London
53. A. Banks (2001).
54. A. Dally, *A Doctor's Story* (London: Macmillan, 1990), pp. 127–132.
55. A. Thorley (2002).
56. H. D. Beckett, 'Heroin, the gentle drug', *New Society*, 49, 877 (1979), 181–182.
57. H. D. Beckett, Interview by Sarah Mars (2001).
58. A. Banks (2001).
59. Senior Civil Servant, DHSS (2001).
60. A. Banks (2001).
61. A. Dally (1990), pp. 127–132.
62. Senior Civil Servant, DHSS (2001).
63. Home Office, MWG (84) 33, 'Power to restrict licences to prescribe under Misuse of Drugs Act 1971' (1984) Private archive.
64. A. Thorley (2002).
65. Senior Civil Servant, DHSS (2001).
66. D. Turner (2002).
67. Senior Civil Servant, DHSS (2001).
68. *Ibid.*
69. *Ibid.*
70. A. Thorley (2002).
71. Senior Civil Servant, DHSS (2001).
72. A. Thorley (2002).
73. Senior Civil Servant, DHSS (2001).
74. ACMD, *Treatment and Rehabilitation*, DHSS (London: HMSO, 1982), p. 59.
75. Medical Working Group on Drug Dependence (1984), p. 3.
76. Senior Civil Servant, DHSS (2001).
77. Senior Civil Servant, DHSS (2001).
78. H. B. Spear (and ed. J. Mott) (2002), p. 279.
79. UK Health Departments, *Drug Misuse and Dependence. Guidelines on Clinical Management* (London: The Stationery Office, 1999), p. 47
80. For example, DHSS, *Better Services for the Mentally Ill* (London: HMSO, 1975) pp. 68–70 and ACMD (1982), pp. 81–86.
81. J. L. Reed, 'Meeting of Doctors Working in London Drug Dependency Treatment Centres, November 25th, 1969 at St Bartholomew's Hospital' [Minutes], Private Archive.
82. H. B. Spear (and ed. J. Mott) (2002), pp. 279–280.
83. ACMD (1982), p. 84.
84. D. Black, 'Medical Working Group on Drug Dependence' (1985), File 16/DAC 28/2, DH Archive, Nelson, Lancashire.
85. D. Mellor, Letter to R. Whitney (10th December 1985), File 16/DAC 28/2, DH Archive, Nelson, Lancashire.
86. J. Patten, Letter to D. Mellor (15th May 1985), File 16/DAC 28/2, Department of Health Archive, Nelson, Lancashire.
87. *Ibid.*
88. R. Witney, Letter to D. Mellor (22nd November 1985), File 16/DAC 28/2, DH Archive, Nelson, Lancashire.

89. For the prescribing figures and a more detailed discussion, see S. G. Mars, *Prescribing and Proscribing: The Public-Private Relationship in the Treatment of Drug Addiction in England, 1970–99* (University of London: PhD Thesis, 2005), pp. 103–104.

90. D. Acheson, Memo to Ms Bateman (30th October 1985), File 16/DAC 28/2, DH Archive, Nelson, Lancashire.

91. D. Black, 'Medical Working Group on Drug Dependence Report on the Feasibility of the Extension of Licensing Restrictions to All Opioid Drugs' Memo to Dr Mason, Dr Oliver and Ms McKessack. (1st October 1985), File 16/DAC 28/2, DH Archive, Nelson, Lancashire.

92. Mellor, D. Letter to R. Whitney (10th December 1985), File 16/DAC 28/2, Department of Health Archive, Nelson, Lancashire.

93. Senior Civil Servant, DHSS (2001).

94. For example, K. Clarke, 'Tackling drugs misuse' (31st October 1983), File DAC 14/4, DH Archive, Nelson, Lancashire.

95. J. Patten, Thames Television (1984).

96. Medical Working Group on Drug Dependence (1984), p. 7.

97. Senior Civil Servant, DHSS (2001).

98. A. Thorley (2002).

99. AIDA, 'Comments on "Guidelines of Good Clinical Practice in the Treatment of Drug Misuse" (DHSS 1984)' (1985) Ref. 44020, DrugScope Library, London.

100. *Ibid.* p. 5.

101. A. Dally, *A Doctor's Story* (London: Macmillan, 1990), pp. 127–132.

102. *Ibid.* p. 130.

103. A. Thorley (2002).

104. ACMD (1982), p. 84.

105. G.V. Stimson and E. Oppenheimer (1982), pp. 215–219.

106. H. B. Spear (and ed. J. Mott) (2002), pp. 235–310.

107. *Ibid.* pp. 245–246.

108. A. Thorley (2002).

109. UK Health Departments (1999).

110. Department of Health (England) and the devolved administrations, *Drug Misuse and Dependence: UK Guidelines on Clinical Management* (London: Department of Health [England], the Scottish Government, Welsh Assembly Government and Northern Ireland Executive, 2007).

111. S. Harrison and W.I.U. Ahmad (2000), pp. 129–146.

112. R. Hartnoll, M.C. Mitcheson, A. Battersby, *et al.*, 'Evaluation of heroin maintenance in controlled trial', *Archives of General Psychiatry*, 37 (1980), 877–884.

113. J. Strang, '"The British System": past, present and future', *International Review of Psychiatry*, 1 (1989), 109–120, p. 116.

114. M. Mitcheson (1994), p. 189.

115. S. M. Shortell, 'Occupational prestige differences within the medical and allied health professions', *Social Science and Medicine*, 8 (1974), 1–9.

116. G. V. Stimson and E. Oppenheimer (1982), pp. 215–219.

117. T. Bewley (2001).

118. Anonymous, 'Journal Interview 36: Conversation with Thomas Bewley', *Addiction*, 90 (1995), 883–892, p. 885.

119. A. Thorley (2002).
120. H. D. Beckett (2001).
121. T. H. Bewley and R. S. Fleminger, 'Staff/patient problems in drug dependence treatment clinics', *Journal of Psychosomatic Research*, 14 (1970), 303–306, p. 306.
122. T. Bewley (2001).
123. J.H. Willis, 'Unacceptable face of private practice: prescription of controlled drugs to addicts' [letter], *British Medical Journal*, 287 (1983), 500.
124. Patient Interview 2 by Sarah Mars (2002).
125. A. Thorley (2002).
126. *Ibid.*
127. V. Berridge, 'AIDS and the rise of the patient? Activist organisation and HIV/AIDS in the UK in the 1980s and 1990s', *Medizin Gesellschaft und Geschichte*, 21 (2002), 109–123.
128. S. Harrison and W. I. U. Ahmad (2000), pp. 129–146.
129. ACMD (1982), pp. 51–62.
130. A. Burr, 'The Piccadilly drug scene', *British Journal of Addiction*, 78, 1 (1983), 5–19.
131. H. B. Spear (and ed. J. Mott) (2002), pp. 218–222.
132. R. Klein, *The New Politics of the National Health Service* (London and New York: Longman, first published 1983, third edition 1995), pp. 243–244.
133. A. Thorley (2002).
134. Senior Civil Servant, DHSS (2001).
135. J. Stanton, 'What shapes vaccine policy? The case of hepatitis B in the UK', *Social History of Medicine*, 7, 3 (1994), 427–446.

4   Ambiguous Justice: The General Medical Council and Dr Ann Dally

1. For example, M. Ashton, 'Doctors at war', *Druglink*, 1, 1 (1986), 13–15 and A. Mold, Heroin. *The Treatment of Addiction in Twentieth-Century Britain* (DeKalb, IL: Northern Illinois University Press, 2008), pp. 97–103.
2. For example, D. Brahams, '"Serious Professional Misconduct" in Relation to Private Treatment of Drug Dependence', *The Lancet*, i (1987), 340–341, p. 341.
3. R. G. Smith, Medical Discipline. *The Professional Conduct Jurisdiction of the General Medical Council, 1858–1990* (Oxford: Oxford University Press, 1994), pp. 1–31.
4. M. Stacey (1992) *Regulating British Medicine: the General Medical Council* (Chichester: John Wiley and Sons, 1992), p. 173.
5. H. B. Spear (and ed. J. Mott), *Heroin Addiction Care and Control: The 'British System' 1916–1984* (London: DrugScope, 2002), p. 64.
6. M. Stacey (1992), pp. 173 and 184.
7. *Ibid.* pp. 181–199.
8. *Ibid.*
9. M. Stacey, 'The General Medical Council and Professional Self-Regulation', in D. Gladstone (ed.), *Regulating Doctors* (London: Institute for the Study of Civil Society, 2000), pp. 28–39.

10. Hansard, House of Lords (14th January 1971), Vol 314, col. 248.
11. R. G. Smith (1994), p. 180.
12. *Ibid*. pp. 103–104, 110.
13. D. Hawks, 'The dimensions of drug dependence in the United Kingdom', in G. Edwards, M. A. H. Russell, D. Hawks et al. (eds.), *Drugs and Drug Dependence* (Farnborough: Saxon House and Lexington, MA: Lexington Books, 1976), pp. 7–10.
14. Interdepartmental Committee on Drug Addiction, The Second Report of the Interdepartmental Committee (London: HMSO, 1965).
15. Hansard, House of Lords (14th January 1971), Vol 314, col. 245.
16. Committee of Inquiry into the Regulation of the Medical Profession, Report of the Committee of Inquiry into the Regulation of the Medical Profession [Merrison Report] (London: HMSO, 1975), p. 87.
17. M. Stacey (1992), pp. 28–39.
18. A. Dally, *A Doctor's Story* (London: Macmillan, 1990), p. 113.
19. R. G. Smith (1994), p. 39.
20. House of Commons Social Services Committee, Fourth Report: Misuse of Drugs with Special Reference to the Treatment and Rehabilitation of Misusers of Hard Drugs, Session 1984–1985 (London: HMSO, 1985), pp. 67–68.
21. Medical Working Group on Drug Dependence, *Guidelines of Good clinical Practice in the Treatment of Drug Misuse* (London: DHSS, 1984).
22. H. B. Spear (and ed. J. Mott) (2002), p. 63.
23. House of Commons Social Services Committee (1985), p. 68.
24. M. Stacey (1992), p. 163.
25. *Ibid*. p. 13.
26. Without GMC registration it is not an offence to practice medicine, only to claim to be a 'registered medical practitioner'.
27. S. Anderson, 'Health professionals and health care systems: the role of the state in the development of community pharmacy in Great Britain 1900 to 1990', National Health Policies in Context Workshop (Bergen, Norway, 27–28th March 2003).
28. M. Stacey (2000), pp. 28–39.
29. R. Cooter (2000), pp. 458–60.
30. Home Office Inspector (2002).
31. *Ibid*.
32. For example, A. Dally (1990), p. 134.
33. *Ibid*. pp. 67–76.
34. R. G. Smith (1994), p. 217.
35. House of Commons Social Services Committee (1985), pp. 167–187.
36. AIDA, 'Management of Addiction', [flier announcing formation of AIDA] (November 1981), File PP/DAL/B/4/1/1/2, Wellcome Library, London.
37. For example, AIDA, 'Drug Addiction: Guidelines and Standards in Management', Pre-publication edition. (February 1982). File PP/DAL/B/4/1/1/5, Wellcome Library, London.
38. A. Dally (1990), pp. 87–98.
39. A. Dally, Letter to N. P. Da Sylva (27th February 1984), File PP/DAL/B/4/1/1/1, Wellcome Library, London.
40. AIDA, Minutes of Meeting held at the Home Office on 29th July, 1982, File PP/DAL/B/4/1/1/4, Wellcome Library, London.

41. GMC, Professional Conduct Committee, Day Two (6th July 1983), Case of Dally, Ann Gwendolen, T A Reed & Co. [transcript], GMC Archive, London, pp. 2 and 53.

42. AIDA, 'Drug Addiction: Guidelines and Standards in Management', Pre-publication edition (February 1982). File PP/DAL/B/4/1/1/5, Wellcome Library, London.

43. D. Black, Letter to A. Dally (19th March 1982), File PP/DAL/B/4/1/1/5, Wellcome Library, London.

44. AIDA, 'Drug Addiction: Guidelines and Standards in Management', Pre-publication edition (February 1982). File PP/DAL/B/4/1/1/5, Wellcome Library, London.

45. D. Black, Letter to A. Dally (19th March 1982), File PP/DAL/B/4/1/1/5, Wellcome Library, London.

46. Diconal is the trade name for the opiate dipipanone combined with the anti-nausea drug cyclizine.

47. AIDA, 'Comments on: Department of Health and Social Security: Treatment and Rehabilitation (HMSO. 1982). Report of the ACMD' (January 1983), File PP/DAL/B/4/1/1/7, Wellcome Library, London, p. 7.

48. GMC Professional Conduct Committee, Day Two (6 July 1983), Case of Dally, Ann Gwendolen, T A Reed & Co. [transcript], GMC Archive, London, pp. 2 and 12.

49. AIDA, 'Meeting, Home Office, 16.9.82' [handwritten notes], File PP/DAL/B/4/1/1/4, Wellcome Library, London.

50. A. Dally, Letter to Dr D. D. P. Rai (17th January 1983), File PP/DAL/B/4/1/1/1 (File 1 of 2), Wellcome Library, London.

51. A. Dally, *A Doctor's Story* (London: Macmillan, 1990).

52. M. Thatcher, Letter to A. Dally (11th August 1983), File PP/DAL/B/5/1/1, Wellcome Library, London.

53. H. B. Spear (and ed. J. Mott) (2002), p. 216.

54. T. H. Bewley, 'Prescribing psychoactive drugs to addicts', *British Medical Journal*, 281 (1980), 497–498, p. 497.

55. The Lancet, 'Drug addiction: British System failing', *The Lancet*, 1 (1982) 83–84, p. 83.

56. ACMD, *Treatment and Rehabilitation*, DHSS (London: HMSO, 1982).

57. A. H. Ghodse, 'Treatment of drug addiction in London', *The Lancet*, 1 (1983), 636–639.

58. A. Dally, 'Drug clinics today', *The Lancet*, 1 (1983), 826.

59. T. Bewley and A. H. Ghodse, 'Unacceptable face of private practice: prescription of controlled drugs to addicts', *British Medical Journal*, 286 (1983), 1876–1877.

60. Ibid.

61. T. Bewley, Interview by Sarah Mars (2001).

62. *British Medical Journal*, 'Doctors for drug addicts', *British Medical Journal*, 286 (1983), 1844.

63. T. Bewley (2001).

64. Ritalin is the trade name for the stimulant methylphenidate.

65. Interdepartmental Committee on Drug Addiction, The Second Report of the Interdepartmental Committee (London: HMSO, 1965), p. 6.

66. H. B. Spear, 'The early years of the "British System" in practice', in J. Strang and M. Gossop (eds.), *Heroin Addiction And Drug Policy: The*

British System (Oxford; New York; Tokyo: Oxford University Press, 1994), pp. 3–28.

67. H. B. Spear (and ed. J. Mott) (2002), p. 145,
68. T. Bewley and A. H. Ghodse (1983), p. 1877.
69. H. B. Spear (and ed. J. Mott) (2002), p. 287.
70. British Medical Journal, 'Doctors for drug addicts', *British Medical Journal*, 286, (1983), 1844.
71. P. Dally, 'Unacceptable face of private practice: prescription of controlled drugs to addicts', [letter], *British Medical Journal*, 287 (1983), 500; H. D. Beckett, 'Prescription of controlled drugs to addicts' [letter], *British Medical Journal*, 287 (1983), 127; E. Stungo, 'Prescription of controlled drugs to addicts' [letter], *British Medical Journal*, 287 (1983), 126–127; R. Hartnoll and R. Lewis, 'Unacceptable face of private practice: prescription of controlled drugs to addicts', [letter], *British Medical Journal*, 287 (1983), 500; A. B. Robertson, 'Prescription of controlled drugs to addicts' [letter], *British Medical Journal*, 287 (1983), 126.
72. A. Dally, Interview by Sarah Mars (2002).
73. A. Dally (1990), p. 100.
74. H. B. Spear (and ed. J. Mott) (2002), p. 287.
75. GMC, Professional Conduct Committee, Day One (5th July 1983), Case of Dally, Ann Gwendolen, T A Reed & Co. [transcript], GMC Archive, London, p. 1.
76. *Ibid*. p. 40.
77. GMC, Professional Conduct Committee, Day Two (6th July 1983), Case of Dally, Ann Gwendolen, T A Reed & Co. [transcript], GMC Archive, London, pp. 2 and 30.
78. Ibid. pp. 2/62–2/63.
79. For example, M. Ashton, 'Doctors at war', *Druglink*, 1 (1986), 13–15; J. Laurance and A. Dally, 'Racketeer or rescuer?', *New Society*, 1256 (1987), 18–19; Anonymous, 'Heroin on the NHS', *New Society*, 1269 (1987), 3; Panorama, BBC1 (1987).
80. A. Dally (1990), p. 134.
81. Home Office Inspector (2002).
82. Dally, A. Letter to M. Thatcher (1st June 1984), File PP/DAL/B/4/1/1/1 (File 2 of 2), Wellcome Library, London.
83. M. Thatcher, Letter to A. Dally (7th June 1984), File PP/DAL/B/4/1/1/1 (File 2 of 2), Wellcome Library, London.
84. A. Dally (1990), pp. 145–149.
85. Home Office Inspector (2002).
86. GMC, Professional Conduct Committee, Day One (9th December 1986), Case of Dally, Ann Gwendolen, T A Reed & Co. [transcript], GMC Archive, London, pp. 1/71–1/72.
87. Privy Council, Appeal No.7 of 1987, Ann Gwendolen Dally v. The General Medical Council, Judgment of the Lords of the Judicial Committee of the Privy Council, Delivered the 14th September 1987, pp. 1–8.
88. *Ibid*. pp. 2–3.
89. GMC, Professional Conduct Committee. Day Two (10th December 1986), Case of Dally, Ann Gwendolen, T A Reed & Co. [transcript], GMC Archive, London, p. 2/7.

90. GMC. Professional Conduct Committee, Day One (9th December, 1986), Case of Dally, Ann Gwendolen, T A Reed & Co. [transcript], GMC Archive, London, p. 1/48.
91. R. G. Smith (1994), pp. 70–77.
92. GMC, Professional Conduct Committee (4th July 1988), Case of Dally, Ann Gwendolen (Resumed case) T A Reed & Co. [transcript], GMC Archive, London. p. 48.
93. Home Office Inspector (2002).
94. R. G. Smith (1994), p. 163.
95. A. Dally (1990), p. 122.
96. *Ibid.* p. 134.
97. Home Office Inspector (2002).
98. GMC, Professional Conduct Committee, Day Two (6th July 1983), Case of Dally, Ann Gwendolen, T A Reed & Co. [transcript], GMC Archive, London, pp. 2/55–2/56.
99. GMC, Professional Conduct Committee, Day Two (6th July 1983), Case of Dally, Ann Gwendolen, T A Reed & Co. [transcript], GMC Archive, London, pp. 2/61–2/63.
100. *Ibid.* p. 2/53.
101. Home Office Inspector (2002).
102. P. Spurgeon, Interview by Sarah Mars (2004).
103. A. Dally (1990), pp. 99–276.
104. M. O'Donnell, 'One man's burden', *British Medical Journal*, 294 (1987), 451.
105. Memorandum from the General Medical Council, House of Commons Social Services Committee, Fourth Report: Misuse of Drugs with Special Reference to the Treatment and Rehabilitation of Misusers of Hard Drugs, Session 1984–85 (London: HMSO, 1985), p. 67.
106. D. Brahams, '"Serious Professional Misconduct" in Relation to Private Treatment of Drug Dependence', *The Lancet*, i, (1987), 340–341, p. 341.
107. A. Dally (1990), p. 196.
108. GMC, Professional Conduct Committee, Day One (9th December 1986), Case of Dally, Ann Gwendolen, T A Reed & Co. [transcript], GMC Archive, London, p. 1/12.
109. GMC, Professional Conduct Committee, Day Three (11th December 1986), Case of Dally, Ann Gwendolen, T A Reed & Co. [transcript], GMC Archive, London, pp. 3/56–3/58.
110. GMC, Professional Conduct Committee, Day One (9th December 1986), Case of Dally, Ann Gwendolen, T A Reed & Co. [transcript], GMC Archive, London, pp. 1/50–1/59.
111. Thomson Directories, Ltd, London Classified Trades and Professions, Telephone Directory 1968 (London: General Post Office, 1968), p. 940.
112. GMC, Professional Conduct Committee, Day Four (9th March 1984), Case of Tarnesby, Herman Peter, T A Reed & Co. [transcript], GMC Archive, London, pp. 16–18.
113. GMC, Professional Conduct Committee, Day Five (10th March 1984) Case of Tarnesby, Herman Peter, T A Reed & Co. [transcript], GMC Archive, London, pp. 88–90.

114. GMC, Professional Conduct Committee. Day One (6th March 1984), Case of Tarnesby, Herman Peter, T. A. Reed & Co. [transcript], GMC Archive, London. unnumbered page, preceding, p. 1.

115. GMC, Professional Conduct Committee, Day Four (9th March 1984), Case of Tarnesby, Herman Peter, T A Reed & Co. [transcript], GMC Archive, London, pp. 1–79.

116. *Ibid*. pp. 11–12.

117. GMC, Professional Conduct Committee, Day Five (10th March 1984) Case of Tarnesby, Herman Peter, T A Reed & Co. [transcript], GMC Archive, London, pp. 17–22.

118. GMC, Professional Conduct Committee, Day Two (7th March 1984), Case of Tarnesby, Herman Peter, T. A. Reed & Co. [transcript], GMC Archive, London, pp. 33–35.

119. GMC, Professional Conduct Committee, Day Four (9th March 1984) Case of Tarnesby, Herman Peter, T A Reed & Co. [transcript], GMC Archive, London, p. 27.

120. GMC, Professional Conduct Committee, Day Five (10th March 1984) Case of Tarnesby, Herman Peter, T A Reed & Co. [transcript], GMC Archive, London, p. 85.

121. GMC, Professional Conduct Committee, Day Two (6th July 1983) Case of Dally, Ann Gwendolen, T A Reed & Co. [transcript], GMC Archive, London, pp. 2/82–2/83.

122. D. Brahams, 'No right of appeal against GMC finding of serious professional misconduct without suspension or erasure', The Lancet, 2 (1983), 979–981, p. 979.

123. *Ibid*. p. 980.

124. GMC, Professional Conduct Committee, Day Five (10th March 1984), Case of Tarnesby, Herman Peter, T A Reed & Co. [transcript], GMC Archive, London, p. 87.

125. GMC. Professional Conduct Committee. Day Four (9th March 1984), Case of Tarnesby, Herman Peter, T. A. Reed & Co. [transcript], GMC Archive, London, p. 2–9.

126. GMC, Professional Conduct Committee (4th July 1988) Case of Dally, Ann Gwendolen (Resumed case) T A Reed & Co. [transcript], GMC Archive, London, p. 48.

127. Medical Working Group on Drug Dependence, Guidelines of Good Practice in the Treatment of Drug Misuse (London: DHSS, 1984).

128. GMC, Professional Conduct and Discipline: Fitness to Practice, edition 61 (London, GMC, 1985).

129. T. H. Bewley (1980), pp. 497–498.

130. P. H. Connell and M. Mitcheson, 'Necessary safeguards when prescribing opioid drugs to addicts: experience of drug dependence clinics in London', *British Medical Journal*, 288 (1984), 767–769.

131. GMC, Professional Conduct Committee, Day Three (8th March 1984), Case of Tarnesby, Herman Peter, T. A. Reed & Co. [transcript], GMC Archive, London, p. 54.

132. GMC. Professional Conduct Committee. Day One, Tuesday 6th March 1984, Case of Tarnesby, Herman Peter, T. A. Reed & Co. [transcript], GMC Archive, London, p. 1/1.

133. See M. Stacey (1992), p. 141.
134. For example, J. Strang, 'Personal View', *British Medical Journal*, 284 (1982), 972.
135. G. V. Stimson and R. Lart, 'The Relationship Between the State and Local Practice in the Development of National Policy on Drugs between 1920 and 1990', in J. Strang and M. Gossop (eds.), *Heroin Addiction And Drug Policy: The British System* (Oxford, New York, Tokyo: Oxford University Press, 1994), pp. 331–341, p. 336.
136. For example, J. Ritchie, 'Drug crazy Britain', *The Sun*, London (17th December 1980), 14–15.
137. A. Dally, Letter to M. Bishop (5th July 1983), File PP/DAL/B/4/1/1/1 (File 1 of 2), Wellcome Library, London.
138. For example, The Lancet, 'Drug dependence in Britain: a critical time', *The Lancet*, 2 (1983), 493–494; *The Lancet*, 'Drug addiction: British System failing', *The Lancet*, 1 (1982), 83–84; M. Honigsbaum, 'The addiction arguments that divide the doctors', Hampstead and Highgate Express, (6th May 1983), 2; Anonymous, 'Doctor Death', *The Listener* (29th July 1982), 22.
139. See J. Merritt, 'Doctors who trade in misery', *Daily Mirror*, (18th February 1982), 7–8; GMC, Professional Conduct Committee, Day Three (8th March 1984), Case of Tarnesby, Herman Peter, T. A. Reed & Co. [transcript], GMC Archive, London. pp. 39–85; GMC, Professional Conduct Committee, Day Four (9th March 1984) Case of Tarnesby, Herman Peter, T A Reed & Co. [transcript], GMC Archive, London. pp. 11–12 and 34–39; GMC, Professional Conduct Committee, Day Five (10th March 1984) Case of Tarnesby, Herman Peter, T A Reed & Co. [transcript], GMC Archive, London, pp. 13–19 and 48–51.
140. Medical Working Group on Drug Dependence, Guidelines of Good Practice in the Treatment of Drug Misuse (London: DHSS, 1984).
141. A. Dally, Letter to S. Perrins (7th March 1983), File PP/DAL/B/4/1/1/1 (File 1 of 2), Wellcome Library, London.
142. I. Munro, Letter to A. Dally (day and month missing, 1983), File PP/DAL/B/4/1/1/1 (File 1 of 2), Wellcome Library, London. P. Chorlton, Letter to A. Dally (20th July 1983), File PP/DAL/B/4/1/1/1 (File 1 of 2), Wellcome Library, London; D. Brahams (1983), pp. 979–981; E. Harbridge, Letter to A. Dally (14th July 1983), File PP/DAL/B/4/1/1/1 (File 1 of 2), Wellcome Library, London.
143. M. O'Donnell, 'One man's burden', *British Medical Journal*, 287 (1983), 990.
144. GMC, Professional Conduct Committee, Day One (9th December, 1986), Case of Dally, Ann Gwendolen, T A Reed & Co. [transcript], GMC Archive, London, p. 1/21.
145. M. Ashton (1986).
146. M. Ashton, 'Doctors at War', *Druglink*, 1(2) (1986), 14–16.
147. R. Baker, 'British and American Conceptions of Medical Ethics, 1847–1947', Anglo-American Medical Relations: Historical Insights Conference, 19th–21st June 2003, The Wellcome Building, London, UK.
148. D. Porter and R. Porter, *Patient's Progress: Doctors and Doctoring in Eighteenth Century England* (Stanford, CA: Stanford University Press, 1989).
149. M. Stacey (1992), p. 57.

150. R. Cooter (2000), pp. 457–460.
151. M. Stacey (1992), pp. 173, 181–199.
152. GMC, Professional Conduct Committee, Day One (9th December 1986), Case of Dally, Ann Gwendolen, T. A. Reed & Co. [transcript], GMC Archive, London, p. 1/47.

5 'Friendly' Visits and 'Evil Men': The Home Office Drugs Inspectorate

1. Home Office Inspector, Interview by Sarah Mars (2002).
2. H. B. Spear (and ed. J. Mott), *Heroin Addiction Care and Control: The 'British System' 1916–1984* (London: DrugScope, 2002), p. 35.
3. Under the Misuse of Drugs Act, 1971.
4. H. B. Spear (and ed. J. Mott), *Heroin Addiction Care and Control: The 'British System' 1916–1984, op. cit.*
5. H. B. Spear placed its origins in 1916 when the Home Secretary authorised a temporary administrative assistant, A. J. Anderson, to inspect records of cocaine supplies which pharmacists had been required to keep from earlier the same year. 1920 was the year of the Dangerous Drugs Act which required pharmacy inspections, and Gerald Rhodes gave that year for the establishment of the Inspectorate, but provided no source, see G. Rhodes, *Inspectorates in British Government. Law Enforcement and Standards of Efficiency*, Royal Institute of Public Administration (London: George Allen and Unwin, 1981), p. 253. Margaret Stacey seemed confused by what she called the Home Office's 'drug squad' the origins of which she placed in the late 1960s, see M. Stacey, *Regulating British Medicine: the General Medical Council* (Chichester: John Wiley and Sons, 1992), p. 32.
6. J. Scullion, Witness Statement (10th March 2003), Document WS 35 00001, The Shipman Inquiry, www.theshipmaninquiry.org.uk.
7. A. Stears, Witness Statement (31st August 1999), Document WS 34 00001, The Shipman Inquiry, www.theshipmaninquiry.org.uk.
8. H. B. Spear (and ed. J. Mott) (2002), p. 35.
9. V. Berridge and G. Edwards, *Opium and the People. Opiate Use in Nineteenth-Century England* (first published 1981 London: Allen Lane; this edition New Haven, CT and London: Yale University Press, 1987), p. 3.
10. *Ibid.* p. 120.
11. V. Berridge, *Opium and the People* (Free Association Books, London: 1999), p. 238.
12. *Ibid.* pp. 241–245.
13. *Ibid.* pp. 247–248.
14. V. Berridge, 'Drugs and social policy: the establishment of drug control in Britain, 1900–30', *British Journal of Addiction*, 79 (1984), 17–29.
15. *Ibid.*
16. V. Berridge, 'War conditions and narcotics control: the passing of Defence of the Realm Act Regulation 40B', *Journal of Social Policy*, 19 (1978), 285–304.
17. H. B. Spear (and ed. J. Mott) (2002), p. 37.
18. The Shipman Inquiry, *Fourth Report – The Regulation of Controlled Drugs in the Community* (London: HMSO, 2004), pp. 45–46.
19. H. B. Spear (and ed. J. Mott) (2002), pp. 35–36.

20. *Ibid.*
21. J. E. Hayzelden, Letter to Chief Officer of Police, Re Home Office Circular 25/1980 Inspection of Pharmacies (12th March 1980), Document WM 17 00062, The Shipman Inquiry, www.theshipmaninquiry.org.uk.
22. Chemist Inspecting Officers continued to patrol pharmacy records throughout the 20th century but with little formal training. They could refer criminal matters for prosecution or liaise with the Royal Pharmaceutical Society or in the case of prescribing questions, the Inspectorate.
23. V. Berridge (1999), pp. 236–239.
24. See Working Party of the Royal College of Psychiatrists and Royal College of Physicians, *Drugs, Dilemmas and Choices* (Gaskell: London, 2000), pp. 206–210.
25. H. B. Spear (and ed. J. Mott) (2002), p. 39.
26. Departmental Committee on Morphine and Heroin Addiction, Report [Rolleston Report] (London: HMSO, 1926), p. 5.
27. See H. B. Spear (and ed. J. Mott) (2002), p. 39.
28. *Ibid.* p. 40.
29. The Shipman Inquiry (2004).
30. H. B. Spear (and ed. J. Mott) (2002), pp. 39–40.
31. K. Leech, 'Bing Spear: appreciations', in H. B. Spear (and ed. J. Mott), *Heroin Addiction Care and Control: The 'British System' 1916–1984* (London: DrugScope, 2002), pp. ix–xi.
32. G. E. Appelbe (2004).
33. For example, M. Mitcheson, Interview by Sarah Mars (2003).
34. For example, J. Mott, Personal Communication (2005).
35. For example, T. Bewley (2001); A. Dally (1990), p. 72.; D. Turner (2002).
36. Home Office Inspector (2002).
37. A. Dally (1990).
38. J. Mott (2005).
39. H. B. Spear (and ed. J. Mott) (2002), p. 310.
40. P. Spurgeon (2004).
41. Home Office Drugs Branch Inspectorate, *Annual Report 1987* (London: Home Office, 1988), Ref. 52150, DrugScope Library, London.
42. *Ibid.*
43. P. Spurgeon (2004).
44. Home Office Drugs Branch Inspectorate, *Annual Report 1985* (London: Home Office, 1986) Ref. 52186, DrugScope Library, London. p. 21.
45. See, for instance, historical introduction to P. Spurgeon, 'Crack and other drugs – myth or menace?' *Looking Ahead 1989 Conference Report* (Swindon: Swindon Crime Concern, 1989), pp. 3–4.
46. P. Spurgeon (2004).
47. A. Macfarlane (2002).
48. *Ibid.*
49. UK Charity Worker, Interview by Sarah Mars (2002).
50. S. L. Hayward, Interview by Sarah Mars (2003).
51. Senior Civil Servant, Home Office, Interview by Sarah Mars (2002).
52. A. Macfarlane (2002).
53. G. E. Appelbe, Interview by Sarah Mars (2004).

54. Hansard, House of Lords (14th January 1971), Vol 314, col. 245.
55. A. Macfarlane, Interview by Sarah Mars (2002).
56. Home Office Inspector, Interview by Sarah Mars (2002).
57. V. Berridge (1984).
58. The Shipman Inquiry (2004).
59. N. Tilley, Interview by Sarah Mars (2002).
60. H. Hampton, Interview by Sarah Mars (2002).
61. The Shipman Inquiry (2004).
62. S. Lutener, Interview by Sarah Mars (2003).
63. G. E. Appelbe (2004).
64. H. B. Spear (and ed. J. Mott) (2002), p. 260.
65. L. Hay, Interview by Sarah Mars (2003).
66. *Ibid.*
67. H. B. Spear (and ed. J. Mott) (2002), pp. 40–41.
68. Home Office Inspector (2002).
69. *Ibid.*
70. Home Office Drugs Branch Inspectorate, *Annual Report 1985* (London: Home Office, 1986) Ref. 52186, DrugScope Library, London, p. 11.
71. J. Scullion, Transcript of Day 149 (27th June 2003), The Shipman Inquiry, www.theshipmaninquiry.org.uk.
72. P. G. Spurgeon, Letter to all Chief Police Officers Re Notes for Chemist Inspecting Officers (17th March 1988), Document WM 17 00163, The Shipman Inquiry, http://www.theshipmaninquiry.org.uk.
73. Home Office Inspector (2002).
74. P. Spurgeon, Interview by Sarah Mars (2004).
75. G. Rhodes (1981), p. 171.
76. M. Weber, *From Max Weber: Essays in Sociology*, H. H. Gerth and C. Wright Mills (eds. and trans.) (New York and Oxford: Oxford University Press, 1946, this edition, 1964), pp. 228–240.
77. H. B. Spear (and ed. J. Mott) (2002), p. 310.
78. D. Musto, 'Bing Spear: appreciations', in H. B. Spear (and ed. J. Mott), *Heroin Addiction Care and Control: The 'British System' 1916–1984* (London: DrugScope, 2002), pp. viii–ix.
79. P. Bean, 'Policing the medical profession: the use of Tribunals', in D. K. Whynes and P. T. Bean (eds.), *Policing and Prescribing. The British System of Drug Control* (Basingstoke: Macmillan, 1991), pp. 60–70.
80. L. Hay (2003).
81. Department of Health and Social Security, Department of Education and Science, Home Office and Manpower Services Commission, *Misuse of Drugs with Special Reference to the Treatment and Rehabilitation of Misusers of Hard Drugs: Government Response to the Fourth Report from the Social Services Committee Session 1984–85* (London: HMSO, 1985), pp. 18–19.
82. A. Macfarlane (2002).
83. Spear wrote sympathetically about Sathananthan in his book and was Chief Inspector when the licence was granted. See H. B. Spear (and ed. J. Mott) (2002), p. 245.
84. A. Macfarlane (2002).

85. K. Sathananthan, Interview by Sarah Mars (2001).
86. Home Office, MWG (84) 33, 'Power to restrict licences to prescribe under Misuse of Drugs Act 1971' (1984) Private archive.
87. House of Commons Social Services Committee (1985), p. xxv.
88. Anonymous, 'Prescription analysis: opioids 1974–1983', Annex to R. Whitney (22nd November 1985).
89. D. Mellor, Letter to R. Whitney (10th December 1985), File 16/DAC 28/2, DH Archive, Nelson, Lancashire.
90. J. Patten, Letter to D. Mellor (15th May 1985), File 16/DAC 28/2, DH Archive, Nelson, Lancashire.
91. A. D. Macfarlane, CGWG (97) 3, Pharmaceutical Diversion and the Prescribing Dimension. Note by the Home Office Drugs Inspectorate' (3rd March 1997), Private archive.
92. *Ibid.*
93. A. Thorley, Interview by Sarah Mars (2002).
94. Thorley, CGWG (97) 17, 'Clinical Guidelines Working Group (attached to CGWG(97)17 'Note by the Home Office on Licensing'), Home Office Licensing: The Options for Encouraging Good Clinical Practice') (May 1997), File 16 TFD-46 Vol 5, DH Archive, Nelson, Lancashire.
95. C. Ford, Interview by Sarah Mars (2002).
96. See A. Macfarlane, 'Changes to the misuse of drugs legislation of controlled drugs prescribe in the treatment of addiction', [Consultation document] (London: Action Against Drugs Unit, Home Office, 17th March 2000).
97. J. Strang. Interview by Sarah Mars (2010).
98. For example, A. Banks, Letter to N. Fowler, MP (1983), Ref. 40117, DrugScope Library, London.
99. For example, B. Beaumont, T. Carnworth, W. Clee, *et al.*, 'Licensing doctors counters the National Strategy', *Druglink*, 15(6) (2001), 25.
100. Departmental Committee on Morphine and Heroin Addiction (1926), p. 8.
101. The law was amended to include these new regulations under the 1926 Dangerous Drugs Act so that the Secretary of State could refer the case of a doctor to a Tribunal if they were providing any of the drugs other than for the purposes of medical treatment.
102. Departmental Committee on Morphine and Heroin Addiction (1926), p. 24.
103. P. Bean (1991), pp. 62–63.
104. R. G. Smith, *Medical Discipline: The Professional Conduct Jurisdiction of the General Medical Council, 1858–1990* (Oxford: Clarendon Press, 1994), pp. 168–169.
105. *Ibid.* p. 65.
106. H. B. Spear (and ed. J. Mott) (2002), p. 62.
107. Hansard, House of Lords (14th January 1971), Vol 314, col. 245.
108. Hansard, House of Lords (14th January 1971), cols 229–30, quoted in P. Bean (1991), p. 64.
109. H. B. Spear (and ed. J. Mott) (2002), p. 62.
110. Hansard (25 March 1970), cols 1457–8, quoted in P. Bean (1991), p. 64.
111. R. G. Smith (1994), p. 39.
112. A. Dally (1990), pp. 149–151.
113. L. Hay (2003).

114. Home Office Inspector (2002).
115. L. Hay (2003).
116. Home Office, *Guidance to Chemist Inspecting Officers* (Home Office, London: 2002), Document WM 17 00388, The Shipman Inquiry, www.theshipmaninquiry.org.uk.
117. Home Office Inspector (2002).
118. H. B. Spear (and ed. J. Mott) (2002), p. 41.
119. GMC Professional Conduct Committee, Day Three (8th March 1984), Case of Tarnesby, Herman Peter, T. A. Reed & Co. [transcript], GMC Archive, London. p. 9.
120. ACMD member, Personal communication (1997).
121. Senior Compliance Officer [Drugs Licensing & Compliance Unit], Personal communication (2010).
122. N. Tilley, Interview by Sarah Mars (2002).
123. R. G. Smith (1994), p. 104.
124. Home Office Inspector (2002).
125. For example, A. Dally (1990), p. 134.
126. Home Office Drugs Inspectorate, 'Irresponsible prescribing enquiries: Investigation, preparation and presentation of evidence' (1983), Private archive.
127. Home Office Inspector (2002).
128. GMC, Professional Conduct Committee, Day Three (8th March 1984), Case of Tarnesby, Herman Peter, T. A. Reed & Co. [transcript], GMC Archive, London. p. 17.
129. Home Office Inspector (2002).
130. See Medical Working Group on Drug Dependence, *Guidelines of Good Practice in the Treatment of Drug Misuse* (London: DHSS, 1984).
131. See Department of Health, Scottish Office Home and Health Department and Welsh Office, *Drug Misuse and Dependence. Guidelines on Clinical Management* (London: HMSO, 1991)
132. See UK Health Departments, *Drug Misuse and Dependence. Guidelines on Clinical Management* (London: The Stationery Office, 1999).
133. For example, T. Bewley and A. H. Ghodse, 'Unacceptable face of private practice: prescription of controlled drugs to addicts', *British Medical Journal*, 286 (1983), 1876–1877.
134. A. Dally (1990), p. 134.
135. T. H. Bewley, Interview by Sarah Mars (2001).
136. A. Dally (1990), p. 72.
137. D. Turner [SCODA], Interview by Sarah Mars (2002).
138. GMC, Professional Conduct Committee, Day One (9th December 1986), Case of Dally, Ann Gwendolen, T A Reed & Co. [transcript], GMC Archive, London. p. 1/50.
139. Senior Civil Servant, DHSS, Interview by Sarah Mars (2001).
140. Home Office Inspector (2002).
141. Home Office Drugs Inspectorate (1983).
142. *Ibid.*
143. *Ibid.*
144. Drugs Branch Inspectorate, *Annual Report 1986* (London: Home Office, 1987), Ref. 49910, DrugScope Library, London.

145. For example, T. Bewley, 'The Illicit Drug Scene', *British Medical Journal*, 2 (1975), 318–320.
146. P. Spurgeon (2004).
147. GMC, Professional Conduct Committee, Day One (9th December 1986), Case of Dally, Ann Gwendolen, T A Reed & Co. [transcript], GMC Archive, London, p. 1/45.
148. *Ibid.* p. 1/45.
149. A. Macfarlane (2002).
150. H. Hudebine, 'Applying cognitive policy analysis to the drug issue: harm reduction and the reversal of the deviantization of drug users in Britain 1985–2000', *Addiction Research and Theory* (forthcoming).
151. *In the Matter of the Misuse of Drugs Act 1971 and In the Matter of Dr John Adrian Garfoot. Minutes of Proceedings at a Misuse of Drugs Tribunal (England and Wales)* (22nd June 1994), W. B. Gurney and Sons [transcript], GMC Archive, London, p. 10.
152. J. Scullion (27th June 2003).
153. GMC, Professional Conduct Committee, Day Three (8th March 1984), Case of Tarnesby, Herman Peter, T. A. Reed & Co. [transcript], GMC Archive, London, pp. 12–15.
154. C. G. Jeffrey, 'Drug control in the United Kingdom', in R. V. Phillipson (ed.), *Modern Trends in Drug Dependence and Alcoholism* (London: Butterworths), pp. 60–74, p. 67.
155. H. B. Spear (and ed. J. Mott) (2002), pp. 42–62.
156. Home Office Inspector (2002).
157. The names of these two doctors have been removed for legal reasons.
158. Home Office, Report of a Tribunal Set Up Under the Misuse of Drugs Act 1971 to enquire into the Conduct of Dr John Adrian Garfoot MB BS MRCS LRCP, 1994 (Dr Garfoot, Private archive), pp. 4–5.
159. Home Office Drugs Branch Inspectorate (1986), p. 12.
160. ACMD (1982), p. 63.
161. House of Commons, Minutes of Evidence Taken before the Social Services Committee (6th February 1985), *Misuse of Drugs with Special Reference to the Treatment and Rehabilitation of Misusers of Hard Drugs, Fourth Report from the Social Services Committee Session 1984–85* (London: HSMO, 1985), pp. 14–21.
162. H. B. Spear, Letter to H. P. Tarnesby (17th June 1982), quoted in GMC, Professional Conduct Committee. Day Three (8th March 1984), Case of Tarnesby, Herman Peter, T. A. Reed & Co. [transcript], GMC Archive, London, p. 20.
163. GMC, Professional Conduct Committee. Day Four (9th March 1984) Case of Tarnesby, Herman Peter, T A Reed & Co. [transcript], GMC Archive, London, p. 77.
164. GMC, Professional Conduct Committee, Day One (9th December 1986), Case of Dally, Ann Gwendolen, T A Reed & Co. [transcript], GMC Archive, London, p. 1/85.
165. GMC, Professional Conduct Committee. Day Four, (9th March 1984), p. 48.
166. GMC, Professional Conduct Committee. Day Three (8th March 1984), Case of Tarnesby, Herman Peter, T. A. Reed & Co. [transcript], GMC Archive, London, p. 17.

167. Home Office Inspector (2002).
168. A. Dally, Interview by Sarah Mars (2002).
169. C. Brewer, Interview by Sarah Mars (2000).
170. P. Spurgeon (2004).
171. GMC, Professional Conduct Committee. Day Four, (9th March 1984) *op. cit.* p. 78.
172. *Ibid.* p. 27.
173. Home Office Inspector (2002).
174. J. Merritt, 'Doctors who trade in misery', *Daily Mirror* (18th February 1982), 7–8.
175. GMC, Professional Conduct Committee. Day Three (8th March 1984) *op. cit.,* p. 3.
176. GMC, Professional Conduct Committee, Day Two (7th March 1984), Case of Tarnesby, Herman Peter, T. A. Reed & Co. [transcript], GMC Archive, London, pp. 39–91.
177. M. Weber (1964), pp. 228–240.
178. G. Rhodes (1981), p. 171.
179. Foucault, M. *The Birth of the Clinic. An Archaeology of Medical Perception*, translated by A. M. Sheridan Smith (first published 1963, Presses Universitaires de France, this edition, New York: Vintage Books, 1994).
180. Foucault, M. *Discipline and Punish. The Birth of the Prison*, translated by A. Sheridan (first published 1975, Edition Gallimard; this edition Harmondsworth, England: Penguin Books, 1991), pp. 257–292.
181. F. Driver, 'Bodies in space. Foucault's account of disciplinary power' in C. Jones and R. Porter (eds.), *Reassessing Foucault: Power, Medicine and the Body* (London and New York: Routledge, 1998, first published 1994), pp. 113–131.
182. D. Armstrong, *Political Anatomy of the Body. Medical Knowledge in the Twentieth Century* (Cambridge: Cambridge University Press, 1983), pp. 7–18.
183. A. Burr, 'The Piccadilly drug scene', *British Journal of Addiction*, 78, 1 (1983), 5–19.
184. H. B. Spear (and ed. J. Mott) (2002), p. 63.
185. S. MacGregor, 'Policy Responses to the Drugs Problem', in *Responding to Drug Misuse. Research and Policy Priorities in Health and Social Care*, ed. S. MacGregor (London and New York: Routledge, 2010), pp. 1–13.
186. J. Strang (2010).
187. The Shipman Inquiry (2004), pp. 239–243.
188. HM Government, *Safer Management of Controlled Drugs. The Government's Response to the Fourth Report of the Shipman Inquiry* (London: 2004), pp. 16–17. http://www.dh.gov.uk/en/Publicationsandstatistics/Publications/PublicationsPolicyAndGuidance/DH_4097904

6 Unifying Hierarchs and Fragmenting Individualists: Three Professional Groups

1. M. Douglas, *Cultural Bias*, Occasional Paper No 35 (London: Royal Anthropological Institute of Great Britain and Northern Ireland, 1978).

2. D. Oldroyd, 'By grid and group divided: Buckland and the English geological community in the Early Nineteenth Century (Essay Review)', *Annals of Science*, 41 (1984), 383–393.

3. M. Thompson, R. Ellis and A. Wildavsky, *Cultural Theory* (Boulder, CO and Oxford: Westview Press, 1990), p. 9.

4. *Ibid.* pp. 5–11.

5. M. Douglas (1978), pp. 5–6.

6. The Ministry of Health became the Department of Health and Social Security in November 1968.

7. J. Mack, Interview by Sarah Mars (2003).

8. H. B. Spear (and ed. J. Mott) (2002), p. 243.

9. 'Meeting of doctors working in drug dependency treatment units in London, 20th January 1970 at St Bartholomew's Hospital', [fragment] Private archive.

10. See J. Willis, Interview by Sarah Mars (2003); M. Mitcheson, Interview by Sarah Mars (2003); J. Mack (2003).

11. T. Bewley, Interview by Sarah Mars (2001).

12. J. Willis (2003).

13. DHSS, 'Drug dependence: clinical conference', [Minutes of meeting] (28th November 1968), Private archive.

14. J. Willis (2003).

15. M. Mitcheson (2003).

16. A. Dally, *A Doctor's Story* (London: Macmillan, 1990), p. 85.

17. *Ibid.* pp. 85–86.

18. H. B. Spear (and ed. J. Mott) (2002), p. 287.

19. ACMD, *Treatment and Rehabilitation*, DHSS (London: HMSO, 1982), pp. 51–62.

20. S. Lloyd Hayward, Interview by Sarah Mars (2003).

21. M. Johnson, Interview by Sarah Mars (2001).

22. M. Johnson (2000).

23. M. Johnson (2001).

24. 'Private Prescribing and Community Safety', [Minutes of meeting] (29th August 1996), Private archive.

25. AIP, 'Minutes of second meeting regarding prescribing and treatment for drug users', [Minutes of meeting] (23rd January 1997), Private archive.

26. C. Brewer, Interview by Sarah Mars (2003).

27. D. Samways, Letter to A. Dally (9th August 1982), File PP/DAL/B/4/1/1/1 (File 1 of 2), Wellcome Library, London.

28. AIP, 'AIP proposed guidelines, 2nd Draft' (31st July 1997), Private archive.

29. AIP, 'Regarding prescribing and treatment for drug users' [Minutes of meeting] (13th March 1997), Private archive.

30. M. Johnson (2000).

31. AIDA, 'Notes of first meeting, Tuesday 24th November, 1981' (January, 1982), File PP/DAL/B/1/1/2, Wellcome Library, London.

32. A. Dally, Letter to Dr D. D. P. Rai (17th January 1983), File PP/DAL/B/4/1/1/1 (File 1 of 2), Wellcome Library, London.

33. GMC, Professional Conduct Committee, Day Two (6th July 1983), Case of Dally, Ann Gwendolen, T A Reed & Co. [transcript], GMC Archive, London. pp. 2/55–2/57.

34. D. D. P. Rai, Letter to A. Dally (30th January 1983), File PP/DAL/B/4/1/1/1 (File 1 of 2), Wellcome Library, London.
35. A. Dally, Letter to S. Perrins (7th March 1983), File PP/DAL/B/4/1/1/1 (File 1 of 2), Wellcome Library, London.
36. T. Bewley (2001).
37. R. Hartnoll, M. C. Mitcheson, A. Battersby, G Brown, M Ellis, P. Fleming, N. Hedley, 'Evaluation of heroin maintenance in controlled trial', *Archives of General Psychiatry*, 37 (1980), 877–884.
38. Prior to British Pharmacopoeias' standardised to metric measures in 1968, the 'grain' was the lowest measure of weight in the Apothecaries system equivalent to just under 65 milligrams.
39. M. Sharpe, Interview by Sarah Mars (2001).
40. T. Bewley (2001).
41. H. D. Beckett, Interview by Sarah Mars (2001).
42. DHSS, 'Heroin dependence: clinical conference, 1st December 1969', [Minutes of meeting] (January 1970), Private archive.
43. H. B. Spear (2002), p. 236.
44. M. Mitcheson (2003).
45. DHSS, 'Heroin dependence: clinical conference', [Minutes of meeting] (18th September 1969), Private archive.
46. *Ibid.*
47. M. Johnson (2001).
48. AIP (13th March 1997).
49. H. D. Beckett (2001).
50. M. Douglas, *Cultural Bias*, Occasional Paper No.35 (London: Royal Anthropological Institute, first published 1978, second impression 1979), p. 21.
51. See Ann Dally's description of the cases in *A Doctor's Story* (London: Macmillan, 1990), pp. 99–281.
52. C. Brewer, Interview by Sarah Mars (2003).
53. C. Brewer, Letter to A. Dally (27th June 1988), File PP/DAL/B/4/1/1/1 (File 2 of 2), Wellcome Library, London.
54. M. Johnson (2001).
55. *Ibid.*
56. See H. B. Spear (and ed. J. Mott) (2002)., pp. 151–176 and ACMD (1982), pp. 97–102.
57. For example, M. Johnson (2000).
58. J. Mack (2003).
59. *Ibid.*
60. M. Douglas (1979), p. 20.
61. M. Mitcheson (2003).
62. DHSS, [Minutes of meeting held on 28th November 1968].
63. T. H. Bewley (2001).
64. P. H. Connell, 'Drug dependence in Great Britain: a challenge to the practice of medicine', in H. Steinberg (ed.), *Scientific Basis of Drug Dependence* (London: J&A Churchill, 1969), pp. 291–299, 297.
65. J. Mack (2003).
66. *Ibid.*
67. *Ibid.*

68. 'Meeting of doctors working in drug dependency treatment units in London, 20th January 1970 at St Bartholomew's Hospital'.

69. A. Dally, Letter to I. Munro (3rd December 1981), File PP/DAL/B/4/1/1/1 (File 1 of 2), Wellcome Library, London

70. C. Brewer (2003).

71. H. D. Beckett (2001).

72. A. Dally, Letter to A. Trebach (3rd December 1982), File PP/DAL/B/4/1/1/1 (File 1 of 2), Wellcome Library, London.

73. A. Dally, Letter to Dr A P Gray (21st November 1986), File PP/DAL/B/4/1/1/1 (File 2 of 2), Wellcome Library, London.

74. M. Johnson (2000).

75. A. Dally, Letter to the Editor, *GP Magazine* (8th April 1983), File PP/DAL/B/4/1/1/1 (File 1 of 2), Wellcome Library, London.

76. A. Dally, Letter to *The Guardian* newspaper (8th April 1983), File PP/DAL/B/4/1/1/1 (File 1 of 2), Wellcome Library, London.

77. A. Dally, Letter to I. Munro (22nd January 1982), File PP/DAL/B/4/1/1/1 (File 1of 2), Wellcome Library, London.

78. AIDA, '*Comments on: Department of Health and Social Security: Treatment and Rehabilitation* (HMSO. 1982). Report of the Advisory Council on the Misuse of Drugs' (January 1983), File PP/DAL/B/4/1/1/7, Wellcome Library, London.

79. AIDA, 'AIDA Comments on "Guidelines of Good Clinical Practice in the Treatment of Drug Misuse" DHSS 1984 (July 1985)', File PP/DAL/B/4/1/1/8, Wellcome Library, London.

80. AIP, 'Private Prescribing and Treatment for Drug Users.' [Minutes of meeting] (10th December 1996), Private archive.

81. S. L. Hayward (2003).

82. AIP (10th December 1996).

83. *Ibid.*

84. M. Johnson (2000).

85. T. Bewley (2001).

86. DHSS (January 1970).

87. P. H. Connell, 'Treatment of drug-dependent patients, 1968–1969', *British Journal of Addiction*, 86 (1991), 913–915.

88. DHSS, [Minutes of meeting held on 18th September 1969].

89. DHSS (January 1970).

90. J. Willis (2003).

91. J. Mack (2003).

92. 'Meeting of doctors working in drug dependency treatment units in London, 20th January 1970 at St Bartholomew's Hospital'.

93. AIDA, 'Drug Addiction: Guidelines and Standards in Management', Pre-publication edition., February 1982, File PP/DAL/B/4/1/1/5, Wellcome Library, London.

94. AIP (31st July 1997).

95. AIDA, 'Minutes of first meeting of the Working Party' (24th November, 1981), File PP/DAL/B/4/1/1/4, Wellcome Library, London.

96. M. Johnson (2001).

97. Medical Working Group on Drug Dependence, *Guidelines of Good Clinical Practice in the Treatment of Drug Misuse* (London: DHSS, 1984).

98. Department of Health, Scottish Office Home and Health Department and Welsh Office. *Drug Misuse and Dependence. Guidelines on Clinical Management.* (London: HMSO, 1991).

99. UK Health Departments, *Drug Misuse and Dependence. Guidelines on Clinical Management* (London: The Stationery Office, 1999).

100. M. Johnson (2001).

101. AIP (10th December 1996).

102. M. Johnson (2000).

103. A. Dally (3rd December 1982).

104. M. Johnson (2001).

105. D. Brahams, 'No right of appeal against GMC finding of serious professional misconduct without suspension or erasure', *The Lancet*, i (1983), 600–601.

106. I. Munro, Letter to J. Sharp (20th March 1987), File PP/DAL/B/4/1/3/2, Wellcome Library, London.

107. I. Munro, Letter to A. Dally (20th January 1982), File PP/DAL/B/1/1/1, Wellcome Library, London.

108. I. Munro, Letter to A. Dally (7th January 1982), File PP/DAL/B/1/1/1, Wellcome Library, London.

109. A. Dally, Letter to M. Bishop (5th July 1983), File PP/DAL/B/4/1/1/1 (File 1 of 2), Wellcome Library, London.

110. A. Dally (1990), pp. 100–102.

111. M. Thatcher, Letter to A. Dally (7th June 1984), File PP/DAL/B/4/1/1/1 (File 2 of 2), Wellcome Library, London.

112. House of Commons Social Services Committee, *Misuse of Drugs with Special Reference to the Treatment and Rehabilitation of Misusers of Hard Drugs. Fourth Report of the Social Services Committee, Session 1984–85* (London: HMSO, 1985), pp. lvi–lvii.

113. AIP (10th December 1996).

114. See Department of Health (England) and the devolved administrations, *Drug Misuse and Dependence: UK Guidelines on Clinical Management* (London: Department of Health [England], the Scottish Government, Welsh Assembly Government and Northern Ireland Executive, 2007).

115. Radio Four, 'File on Four' (1997).

116. J. L. Reed, "Meeting of Doctors Working in London Drug Dependency Treatment Centres, November 25th, 1969 at St Bartholomew's Hospital"' [Minutes of meeting], Private Archive.

117. ACMD (1982), pp. 51–62.

118. Medical Working Group on Drug Dependence (1984).

119. UK Health Departments (1999).

120. For example, General Medical Council, Professional Conduct Committee, Day One (9th December 1986), Case of Dally, Ann Gwendolen, T A Reed & Co. [transcript], GMC Archive, London. p. 1/12; and 'In the Matter of the Misuse of Drugs Act 1971 and in the Matter of Dr John Adrian Garfoot', Day One. [Minutes of proceedings at a Misuse of Drugs Tribunal, Wednesday 22nd June 1994], W B Gurney and Sons, London. Private archive of Dr Garfoot, p. 10.

121. DHSS (January 1970).

122. H. B. Spear (and ed. J. Mott) (2002), pp. 236–237.

123. P. H. Connell (1991), pp. 913–915.
124. GMC, Professional Conduct Committee, Day Two [(6 July 1983), Case of Dally, Ann Gwendolen, T A Reed & Co. [transcript], GMC Archive, London., GMC Archive, London, pp. 2/11–2/12.
125. G. V. Stimson and R. Lart, 'The relationship between the state and local practice in the development of national policy on drugs between 1920 and 1990', in J. Strang and M. Gossop (eds.), *Heroin Addiction And Drug Policy: The British System* (Oxford, New York, Tokyo: Oxford University Press, 1994), pp. 331–341, 336.
126. J. Strang, 'Personal view', *British Medical Journal*, 284 (1982), 972.
127. A. Dally, 'Opiate addiction and the independent doctor', Letter to the Editor, *British Medical Journal* (2nd April 1982) File PP/DAL/B/4/2/1, Wellcome Library, London.
128. A. Dally, Letter to the Editor, *The Lancet*, for publication (12th January 1982), File PP/DAL/B/4/1/1/1, Wellcome Library, London.
129. For example, A. Dally, Letter to *The Times* (27th January 1981), File PP/DAL/B/4/1/1/1 (File 1 of 2), Wellcome Library, London.
130. AIDA (February 1982), p. 1; and, for example, A. Dally (8th April 1983).
131. A. Dally, Letter to T. Hillier (20th September 1982), File PP/DAL/B/4/1/1/1 (File 1 of 2), Wellcome Library, London.
132. M. Mitcheson (2003).
133. *Ibid.*
134. Home Office Inspector, Interview by Sarah Mars (2002).
135. *Ibid.*
136. *Ibid.*
137. *Ibid.*
138. C. Fazey, Interview by Sarah Mars (2002).
139. H. B. Spear, Letter to A. Dally (1st February 1982), File PP/DAL/B/4/2/1, Wellcome Library, London.
140. H. B. Spear, Letter to Mr Emery (17th May 1987), File PP/DAL/B/4/1/3/2, Wellcome Library, London.
141. H. B. Spear (and ed. J. Mott) (2002), p. 310.
142. *Ibid.* p. 287.
143. D. Black, Letter to A. Dally (19th March 1982), File PP/DAL/B/4/1/1/5, Wellcome Library, London.
144. C. Fazey (2002).
145. J. Strang, Personal communication (2000).
146. K. Sathananthan, Interview by Sarah Mars (2001).
147. 'Private Prescribing and Community Safety' (29th August 1996).
148. S. L. Hayward (2003).
149. 'Private Prescribing and Community Safety' (29th August 1996).
150. D. Samways (9th August 1982).
151. J. Willis (2003).
152. M. Mitcheson (2003).
153. T. Bewley and A. H. Ghodse (1983), pp. 1876–1877.
154. T. H. Bewley (2001).
155. A. Dally (1990), p. 141.
156. Consultant Psychiatrist, Interview by Sarah Mars (2001).
157. J. Mack (2003).

158. AIP (10th December 1996).
159. AIP (13th March 1997).
160. M. Johnson (2001).
161. J. Mack (2003).
162. A. Banks and T. A. N. Waller, *Drug Addiction and Polydrug Use: The Role of the General Practitioner* (London: ISDD, 1983).
163. Senior Civil Servant, DHSS (2001).
164. A. Banks, Letter to A. Dally (15th March 1984), File PP/DAL/B/5/1/2, Wellcome Library, London.
165. D. Samways (9th August 1982).
166. D. Black (19th March 1982).
167. B. Jarman, Letter to A. Dally (7th April 1982), File PP/DAL/B/4/2/1, Wellcome Library, London.
168. A. Dally, Letter to S. Openshaw (11th February 1982), File PP/DAL/B/1/1/1, Wellcome Library, London.
169. M. Johnson (2000).
170. M. Johnson (2001).
171. *Ibid.*
172. For example, T. H. Bewley (2001).
173. J. Willis (2003).
174. H. D. Beckett (2001).
175. H. B. Spear (and J. Mott ed.) (2002), p. 192.
176. Interdepartmental Committee on Drug Addiction, *The Second Report of the Interdepartmental Committee* (London: HMSO, 1965).
177. M. Gluckman, *Custom and Conflict in Africa* (Oxford: Basil Blackwell, first published 1956, this edition 1982).

7 Guidelines and the Licensing Question

1. V. Berridge, *AIDS in the UK. The Making of Policy, 1981–1994* (Oxford: Oxford University Press, 1996), pp. 92–96.
2. UK Health Departments, *Drug Misuse and Dependence. Guidelines on Clinical Management* (London: The Stationery Office, 1999).
3. Department of Health, 'Government to improve care for drug-misusers – New Guidelines for Doctors', Press release reference 1999/0220 (12th April 1999), www.dh.gov.uk.
4. Department of Health, 'Paper CGWG(97)42 Recommendations to accompany clinical guidelines' (October 1997), Private archive.
5. M. Davies Memo to K. Jarvie, 'Drug Misuse Clinical Guidelines – Leak of Working Papers' (1st July 1998), File 16/DRU 323/22 Vol 1, Department of Health Archive, Nelson, Lancashire.
6. D. Curson, 'Private treatment of alcohol and drug problems in Britain', *British Journal of Addiction*, 86 (1991), 9–11.
7. The author worked for the British Medical Association's Clinical Audit Working Group in the early 1990s.
8. For instance, the Royal College of Psychiatrists' College Research Unit, reliant mainly on outside project funding, received a number of grants from the Department of Health to develop clinical guidelines.

9. Medical Working Group on Drug Dependence, *Guidelines of Good Practice in the Treatment of Drug Misuse* (London: DHSS, 1984).
10. Department of Health, Scottish Office Home and Health Department and Welsh Office. *Drug Misuse and Dependence. Guidelines on Clinical Management.* (London: HMSO, 1991).
11. UK Health Departments, *Drug Misuse and Dependence. Guidelines on Clinical Management* (London: The Stationery Office, 1999).
12. R. Klein, *The New Politics of the National Health Service* (first published 1983, London and New York: Longman; fourth edition, Harlow: Prentice Hall, 2001).
13. UK Health Departments, *Drug Misuse and Dependence*, pp. 9–15, .
14. A. Mold and V. Berridge, *Voluntary Action and Illegal Drugs. Health and Society in Britain Since the 1960s* (Houndsmill, Basingstoke: Palgrave Macmillan, 2010), p. 149.
15. Department of Health (England) and the devolved administrations, *Drug Misuse and Dependence: UK Guidelines on Clinical Management* (London: Department of Health [England], the Scottish Government, Welsh Assembly Government and Northern Ireland Executive, 2007).
16. The Task Force to Review Services for Drug Misusers, *Report of an Independent Review of Drug Treatment Services in England* (London: The Stationery Office, Department of Health, 1996).
17. J. Strang, Letter to J. Polkinghorne (29th November 1995), File 16/TFD-45 Vol 1, DH Archive, Nelson, Lancashire.
18. The Task Force to Review Services for Drug Misusers, p. 67.
19. Sir Kenneth Calman, Letter to Working Group Members (5th November 1996), Private archive.
20. L. Gruer, Interview by Sarah Mars (2003).
21. S. Mars and V. Berridge, 'Social Science Small Grant End of Project Report Ref No: SGS/00742/G, The impact of drug user patient groups ('user groups') on UK drug treatment policy since the 1970s.' [unpublished] (2002).
22. Anonymous, 'Clinical Guidelines Working Group – Note of the meeting of 16th March 1998', File 16 DRU 323/12 Vol 1. DH Archive, Nelson, Lancashire.
23. AIP, 'Private prescribing and treatment for drug users', [Minutes of meeting] (13th March 1997), Private archive.
24. AIP, 'Private prescribing and treatment for drug users', [Minutes of meeting] (10th December 1996), Private archive.
25. J. Strang, J. Bearn, M. Gossop, 'Opiate detoxification under anaesthesia', *British Medical Journal*, 315 (1997), 1249–1250.
26. AIP (13th March 1997).
27. D. Curson (1991), pp. 9–10.
28. A. Thorley, Interview by Sarah Mars (2002).
29. For example, J. Strang, J. Sheridan and N. Barber, 'Prescribing injectable and oral methadone to opiate addicts: results from the 1995 national postal survey of community pharmacies in England and Wales', *British Medical Journal*, 313 (1996), 270–272.
30. In addition to Professor Strang's published criticisms of private prescribing, he has also appeared as an expert witness to give evidence against private

practitioners in a number of disciplinary hearings by the GMC and Home Office Drug Tribunals.

31. See Department of Health (England) and the devolved administrations, *Drug Misuse and Dependence.*
32. Department of Health, 'Clinical Guidelines Working Group Note of meeting held on Friday 7th November 1997 at Waterbridge House' (Undated). Private Archive.
33. C. Ford (2002).
34. C. Ford, Interview by Sarah Mars (2002).
35. *Ibid.*
36. *Ibid.*
37. L. Gruer (2003).
38. British Medical Association. General Medical Services Committee, *Core Services: Taking The Initiative* (London: British Medical Association, 1996).
39. British Medical Association. General Medical Services Committee, *Core Services.*
40. The author was researcher to the British Medical Association's Working Party on Drug Misuse (1995–1997).
41. J. Strang, Interview by Sarah Mars (2002).
42. L. Gruer (2003).
43. S. Mars, 'Peer pressure and imposed consensus: The making of the 1984 "guidelines of good clinical practice in the treatment of drug misuse"', in V. Berridge (ed.), *Making Health Policy: Networks in Research and Policy After 1945* (Amsterdam: Rodopi, 2005), pp. 149–182.
44. J. Strang (2002).
45. Department of Health. CGWG (97) 26, 'Private practice and the prescribing of controlled drugs' (June 1997), Private archive.
46. C. Ford (2002).
47. UK Health Departments (1999), p. 7.
48. *Ibid.* p. 31.
49. Interdepartmental Committee on Drug Addiction, *Drug Addiction. The Second Report of the Interdepartmental Committee* [second Brain Report], Ministry of Health, Scottish Home and Health Department (London: HMSO, 1965), p. 9.
50. UK Health Departments (1999), p. 63.
51. *Ibid.* pp. 42–43.
52. A. Uchtenhagen, F. Gutzwiller and A. Dobler-Mikola, *Programme for a Medical Prescription of Narcotics: Final Report of the Research Representatives. Summary of the Synthesis Report* (Zurich: University of Zurich, 1997).
53. UK Health Departments (1999), p. 57.
54. *Ibid.* p. 55.
55. *Ibid.* p.xiv.
56. ACMD, *AIDS and Drug Misuse Update*, Department of Health (London: HMSO, 1993), p. 53.
57. Macfarlane, A. 'Changes to the misuse of drugs legislation of controlled drugs prescribe in the treatment of addiction', [Consultation document] (London: Action Against Drugs Unit, Home Office, 17th March 2000).
58. 'Shared care' is described by the 1999 *Guidelines* as 'a model that can be applied to any close co-operative work between agencies or services, which

directly improves the treatment of the individual drug misuser. It most often involves arrangements between specialist and general practitioner services', p. 10.

59. *In the Matter of the Misuse of Drugs Act 1971 and In the Matter of Dr John Adrian Garfoot. Minutes of Proceedings at a Misuse of Drugs Tribunal* (England and Wales) (22nd June 1994), W. B. Gurney and Sons [transcript], GMC Archive, London, pp. 42–58.
60. Department of Health, Draft sent to Clinical Guidelines Working Group on 5th March 1998 (undated) Private archive.
61. Department of Health, Draft sent to Clinical Guidelines Working Group on 3rd July 1998 (undated) Private archive.
62. Anonymous, 'Clinical Guidelines Working Group – Note of the meeting of 16th March 1998', File16 DRU 323/12 Vol 1. DH Archive, Nelson, Lancashire.
63. W. Clee, Letter to M. Davies (30th July 1998), Private archive.
64. C. Ford, Letter to M. Davies (18th July 1998), Private archive.
65. For example, J. Bury, Letter to F. Pink (12th October 1998), Private archive.
66. Working Group member, Interview by Sarah Mars (2002).
67. M. Davies, (1998).
68. Working Group member, Interview by Sarah Mars (2002).
69. K. McElrath 'Heroin Use in Northern Ireland: A Qualitative Study Into Heroin Users' Lifestyles, Experiences And Risk Behaviours (1997–1999)' (Belfast: Department of Health, Social Services and Public Safety, 2001).
70. D. Patterson, Letter to M. Davies (8th July 1998) File 16 DRU 323/12 Vol 1, DH Archive, Nelson, Lancashire.
71. L. Gruer (2003).
72. Clinical Guidelines Working Group, Draft guidelines (3rd July 1998) [draft B].
73. Anonymous, 'Clinical Guidelines – Action Points 22nd July '98', File 16, DRU 323/12 Vol 1, DH Archive, Nelson, Lancashire.
74. B. Beaumont, T. Carnworth, W. Clee, *et al.*, 'Licensing doctors counters the National Strategy', *Druglink*, 15(6) (2001), 25.
75. AIDA, *AIDA Comments on Guidelines of Good Clinical Practice in the Treatment of Drug Misuse DHSS 1984* (July 1985), File PP/DAL/B/4/1/1/8, Wellcome Library, London.
76. C. Ford (2002).
77. A. Dally, Letter to I. Munro (22nd January 1982), File PP/DAL/B/4/1/1/1 (File 1of 2), Wellcome Library, London.
78. C. Ford, Facsmile to D. Raistrick Re DoH Injecting Subgroup (1st June 1997). Private archive.
79. C. Ford (2002).
80. C. Ford, Letter to J. Strang, C. Gerada, M. Farrell and M. Davies 'Re: Clinical Guidelines Working Group – Notes of the meeting of 16th March' (24th May 1998), Private archive.
81. C. Ford (2002).
82. see A. Dally, *A Doctor's Story* (London: Macmillan, 1990).
83. C. Ford (2002).
84. ACMD. *Treatment and Rehabilitation*, DHSS (London: HMSO, 1982), pp. 57–60.

85. A. D. Macfarlane, CGWG (97) 17, 'Clinical Guidelines Working Group. Note by the Home Office on Licensing', Home Office (8th May 1997), Private archive.
86. Medical Working Group on Drug Dependence (1984), p. 3.
87. UK Health Departments (1999), p.xv.
88. J. L. Reed, 'Meeting of Doctors Working in London Drug Dependency Treatment Centres, November 25th, 1969 at St Bartholomew's Hospital' [Minutes], Private archive.
89. House of Commons Social Services Committee, *Misuse of Drugs with Special Reference to the Treatment and Rehabilitation of Misusers of Hard Drugs. Fourth Report of the Social Services Committee, Session 1984–85* (London: HMSO, 1985), p. xxv.
90. M. Mitcheson, Interview by Sarah Mars (2003).
91. ACMD (1982).
92. ACMD, *Prevention* (London: HMSO, 1984).
93. D. Mellor, Letter to R. Whitney (10th December 1985), File 16/DAC 28/2, DH Archive, Nelson, Lancashire.
94. J. Patten, Letter to D. Mellor (15th May 1985), File 16/DAC 28/2, DH Archive, Nelson, Lancashire.
95. A. D. Macfarlane, CGWG (97) 3, 'Pharmaceutical Diversion and the Prescribing Dimension. Note by the Home Office Drugs Inspectorate' (3rd March 1997), Private archive.
96. *Ibid.*
97. A. Thorley (2002).
98. A. Thorley, CGWG (97) 17, 'Clinical Guidelines Working Group (attached to CGWG(97)17 'Note by the Home Office on Licensing'), Home Office Licensing: The Options for Encouraging Good Clinical Practice') (May 1997), File 16 TFD-46 Vol 5, DH Archive, Nelson, Lancashire.
99. T. Waller, Interview by Sarah Mars (2001).
100. J. Strang, 'Draft outline for conclusions and recommendations about the prescribing of tablets and ampoules of methadone' (undated; attached to J. Strang, Letter to J. Polkinghorne, 29th November 1995), File 16/TFD-45 Vol 1, DH Archive, Nelson, Lancashire.
101. J. Strang (1995).
102. Task Force to Review Services for Drug Misusers (1996), p. 67.
103. For example, Doctor 010, Interview by Sarah Mars (2000).
104. A. D. Macfarlane (1997).
105. A. Thorley, 'Clinical Guidelines Working Group (attached to CGWG(97)17 'Note by the Home Office on Licensing'), Home Office Licensing: The Options for Encouraging Good Clinical Practice') (8th May 1997), File 16 TFD-46 Vol 5, DH Archive, Nelson, Lancashire.
106. J. Strang, J. Sheridan and N. Barber (1996), 270–272.
107. AIP (13th March 1997).
108. Doctor 011, Interview by Sarah Mars (2001).
109. Doctor 005, Interview by Sarah Mars (2000).
110. For example, A. Thorley, CGWG (97) 5, 'Approaches to the definition of a specialist' (Feb 26th 1997), Private archive.
111. Department of Health. CGWG (97) 26, 'Private practice and the prescribing of controlled drugs' (June 1997), Private archive; A. D. Macfarlane

(8th May 1997); Department of Health, 'Meeting to discuss Misuse of Drugs regulations and private practitioners' (June 1997) Private archive.

112. UK Health Departments (1999), p. 9.
113. M. Johnson, Interview by Sarah Mars (2001).
114. Department of Health (June 1997).
115. Clinical Guidelines Working Group, 'Prescribing and Licensing Regulations' [attached to minutes of 16th March 1998], Department of Health (1998) File 16/DRU 323/12 Vol 1, Department of Health Archive, Nelson, Lancashire.
116. A. Macfarlane (17th March 2000).
117. *Ibid.*
118. For example, A. Banks, Letter to N. Fowler, MP (1983), Ref. 40117, DrugScope Library, London; B. Beaumont, T. Carnworth., W. Clee, *et al.* (2001), p. 25; B. Beaumont, T. Carnworth, W. Clee, *et al.*, 'Licensing doctors counters the National Strategy', *Druglink*, 15(6) (2001), 25.
119. L. Gruer (2003).
120. Senior Civil Servant, DHSS, Interview by Sarah Mars (2001).
121. H. B. Spear (and ed. J. Mott) (2002), pp. 279–283.
122. S. Butler, 'The making of the Methadone Protocol: the Irish system?', *Drugs: Education, Prevention and Policy*, 9, 4 (2002), 311–324.

Conclusion

1. The Telegraph, 'MPs to back Blunkett on prescription heroin', *The Telegraph*, 19th May 2002.
2. A. Travis, 'Coalition shelves plans for "abstinence based" drug strategy.' *The Guardian*, 8th December 2010.
3. V. Berridge, *AIDS in the UK. The Making of Policy, 1981–1994* (Oxford: Oxford University Press, 1996), pp. 90–96.
4. For example, J. L. Reed, "Meeting of Doctors Working in London Drug Dependency Treatment Centres, November 25th, 1969 at St Bartholomew's Hospital" [Minutes], Private archive.; and T. Bewley, 'The Illicit drug scene', *British Medical Journal*, 2 (1975), 318–320.
5. Of the remaining five, one had retired, one had left medical practice for unknown reasons and one had limited his practice to the least controversial of treatments, phasing out injectable medication. Two did not respond to enquiries.
6. M. Johnson, Personal communication, 2011.
7. C. E. Lindblom 'The science of "muddling through"', *Public Administration Review*, 19, 2 (1959), 79–88.
8. V. Berridge, 'Issue network versus producer network? ASH, the Tobacco Products Research Trust and UK smoking policy', in V. Berridge (ed.), *Making Health Policy: Networks in Research and Policy After 1945* (Amsterdam: Rodopi, 2005), pp. 101–124.
9. M. Johnson, Interview by Sarah Mars (2000).
10. M. Johnson, Interview by Sarah Mars (2001).
11. For example, H. D. Beckett, Interview by Sarah Mars (2001).

12. Department of Health and Social Security, 'Heroin dependence: clinical conference, 1st December 1969', [Minutes of meeting] (January 1970), Private archive.
13. Patient 002, Interview by Sarah Mars (2002).
14. For example, Advisory Council on the Misuse of Drugs, *Treatment and Rehabilitation*, DHSS (London: HMSO, 1982), p. 54.
15. For example, A. B. Robertson, 'Prescription of controlled drugs to addicts' [letter], *British Medical Journal*, 287 (1983), 126.
16. A. Dally, *A Doctor's Story* (London: Macmillan, 1990).
17. D. Turner, Interview by Sarah Mars (2002).
18. S. Mars and V. Berridge, 'Social Science Small Grant End of Project Report Ref No: SGS/00742/G, The impact of drug user patient groups ('user groups') on UK drug treatment policy since the 1970s.' [unpublished] (2002).
19. A. Mold and V. Berridge, *Voluntary Action and Illegal Drugs. Health and Society in Britain Since the 1960s* (Houndsmill, Basingstoke: Palgrave Macmillan, 2010), pp. 163–164.
20. See National Treatment Agency for Substance Misuse, *Prescribing Services for Drug Misuse* (London: NTASM, 2003), part of which is authored by patient activist Bill Nelles.
21. A. Mold and V. Berridge (2010), p. 149.
22. Senior Home Office Official, Interview by Sarah Mars (2002).
23. Doctor 005, Interview by Sarah Mars (2000)
24. G. J. Stigler, 'The theory of economic regulation', *Bell Journal of Economics and Management Science*, 2, 1 (1971), 3–21.

Bibliography

Due to the contemporary nature of the debate, all 'secondary' sources also have the potential to be primary sources, being created in or around the period studied, so they are not separated here. Documents listed as in a 'private archive' are those either loaned or given to the author by interviewees during the research project and not held in formal archives. Most of the individual sources of these private documents are not named for reasons of confidentiality.

Archival sources

Acheson, D. Memo to Ms Bateman (30th October 1985), File 16/DAC 28/2, Department of Health Archive, Nelson, Lancashire.

Advisory Council on the Misuse of Drugs, ACMD (82), 'Second meeting minutes' (13th July 1982), File MDS/1/3 Vol 3, Department of Health Archive, Nelson, Lancashire.

Advisory Council on the Misuse of Drugs, 'Report on Treatment and Rehabilitation. Draft Submission' (November 1983), File DAC 14 Vol 4, Department of Health Archive, Nelson, Lancashire.

Anonymous, *In the Matter of the Misuse of Drugs Act 1971 and In the Matter of Dr John Adrian Garfoot. Minutes of Proceedings at a Misuse of Drugs Tribunal (England and Wales)* (22nd June 1994), W. B. Gurney and Sons [transcript]. Private archive of Dr Garfoot.

Anonymous, 'Clinical Guidelines – action points' (22nd July '98), File 16/DRU 323/12 Vol 1, Department of Health Archive, Nelson, Lancashire.

Anonymous, 'Clinical Guidelines Working Group – note of the meeting of 16th March 1998', File 16 DRU 323/12 Vol 1, Department of Health Archive, Nelson, Lancashire.

Anonymous, 'Meeting of doctors working in drug dependency treatment units in London, 20th January 1970 at St Bartholomew's Hospital', [fragment] Private archive.

Anonymous, 'Note of the one-day medical conference convened at the DHSS to discuss the medical response to the ACMD report on Treatment and Rehabilitation on 28th January 1983' (undated), File DAC 28, Department of Health Archive, Nelson, Lancashire.

Anonymous, 'Prescription analysis: opioids 1974–1983', Annex to R. Witney (22nd November 1985) File 16/DAC 28/2, Department of Health Archive, Nelson, Lancashire.

Anonymous, 'Private prescribing and community safety', [Minutes of meeting] (29th August 1996), Private archive.

Anonymous, Draft reply to Home Secretary from Norman Fowler (undated), File DAC 14 Vol 4, Department of Health Archive, Nelson, Lancashire.

Association of Independent Doctors in Addiction, 'Notes of first meeting, Tuesday 24th November, 1981' (January 1982), File PP/DAL/B/1/1/2, Wellcome Library, London.

Association of Independent Doctors in Addiction, 'Management of addiction', [flier announcing formation of AIDA] (November 1981), File PP/DAL/B/4/1/1/2, Wellcome Library, London.

Association of Independent Doctors in Addiction, 'Minutes of first meeting of the Working Party' (24th November 1981), File PP/DAL/B/4/1/1/4, Wellcome Library, London.

Association of Independent Doctors in Addiction, TRWG (82) 10, 'Drug addiction: guidelines and standards in management', Pre-publication edition (February 1982), File DAC 7, Department of Health Archive, Nelson, Lancashire.

Association of Independent Doctors in Addiction, 'Comments on: Department of Health and Social Security: Treatment and Rehabilitation (HMSO 1982). Report of the Advisory Council on the Misuse of Drugs' (January 1983), File PP/DAL/B/4/1/1/7, Wellcome Library, London.

Association of Independent Doctors in Addiction, 'Drug addiction: guidelines and standards in management', Pre-publication edition (February 1982), File PP/DAL/B/4/1/1/5, Wellcome Library, London.

Association of Independent Doctors in Addiction, 'AIDA Comments on "Guidelines of Good Clinical Practice in the Treatment of Drug Misuse" DHSS 1984' (July 1985). File PP/DAL/B/4/1/1/8, Wellcome Library, London.

Association of Independent Doctors in Addiction, 'Comments on "Guidelines of Good Clinical Practice in the Treatment of Drug Misuse" (DHSS 1984)' (1985) Ref. 44020, DrugScope Library, London.

Association of Independent Doctors in Addiction, 'Meeting, Home Office, 16.9.82' [handwritten notes], File PP/DAL/B/4/1/1/4, Wellcome Library, London.

Association of Independent Doctors in Addiction, 'Minutes of meeting held at the Home Office on 29th July, 1982', File PP/DAL/B/4/1/1/4, Wellcome Library, London.

Association of Independent Prescribers, 'Private prescribing and treatment for drug users', [Minutes of meeting] (10th December 1996), Private archive.

Association of Independent Prescribers, 'Minutes of second meeting regarding prescribing and treatment for drug users' (23rd January 1997), Private archive.

Association of Independent Prescribers, 'Minutes of the third meeting regarding prescribing and treatment for drug users' (13th March 1997), Private archive.

Association of Independent Prescribers, 'AIP proposed guidelines, 2nd Draft' (31st July 1997), Private archive.

Association of Independent Prescribers, 'Regarding prescribing and treatment for drug users' [Minutes of meeting] (13th March 1997), Private archive.

Banks, A. Letter to N. Fowler, MP (1983), Ref 40117, DrugScope Library, London

Banks, A. Letter to A. Dally (15th March 1984), File PP/DAL/B/5/1/2, Wellcome Library, London.

Beckett, H. D. Letter to A. Dally (17th June 1983), File PP/DAL/B/4/2/1, Wellcome, Library, London.

Black, D. Letter to A. Dally (19th March 1982), File PP/DAL/B/4/1/1/5, Wellcome Library, London.

Black, D. 'Medical Working Group on Drug Dependence' (1985), File 16/DAC 28/2, Department of Health Archive, Nelson, Lancashire.

Black, D. 'Medical Working Group on Drug Dependence Report on the Feasibility of the Extension of Licensing Restrictions to All Opioid Drugs' Memo to Dr Mason, Dr Oliver and Ms McKessack. (1st October 1985), File 16/DAC 28/2, DH Archive, Nelson, Lancashire.

Blythe, A. M. Memo to J. M. Rogers (19th February 1982), File 16/DAC 28 Vol 2, Department of Health Archive, Nelson, Lancashire.

Blythe, A. M. Memo to M. Moodie. 'Services for drug misusers. Advisory Council on the Misuse of Drugs' (5th February 1982), File DAC 7, Department of Health Archive, Nelson, Lancashire.

Brewer, C. Letter to A. Dally (27th June 1988), File PP/DAL/B/4/1/1/1 (File 2 of 2), Wellcome Library, London.

Bury, J. Letter to F. Pink (12th October 1998), Private archive.

Calman, K. Letter to Working Group Members (5th November 1996), Private archive.

Chorlton, P. Letter to A. Dally (20th July 1983), File PP/DAL/B/4/1/1/1 (File 1 of 2), Wellcome Library, London.

Clarke, K. 'Tackling drugs misuse' (31st October 1983), File DAC 14/4, Department of Health Archive, Nelson, Lancashire.

Clarke, K. Memo to Secretary of State (31st October 1983), File DAC 14 Vol 4, Department of Health Archive, Nelson, Lancashire.

Clee, W. Letter to M. Davies (30th July 1998), Private archive.

Clinical Guidelines Working Group, 'Prescribing and licensing regulations' [attached to minutes of 16th March 1998], Department of Health (1998), File 16/DRU 323/12 Vol 1, Department of Health Archive, Nelson, Lancashire.

Dally, A. 'National resources for drug abuse', Letter to *The Times* (27th January 1981), File PP/DAL/B/4/1/1/1 (File 1 of 2), Wellcome Library, London.

Dally, A. Letter to *The Times* (27th January 1981), File PP/DAL/B/4/1/1/1 (File 1 of 2), Wellcome Library, London.

Dally, A. Letter to I. Munro (3rd December 1981), File PP/DAL/B/4/1/1/1 (File 1 of 2), Wellcome Library, London.

Dally, A. 'Opiate addiction and the independent doctor', Letter to the Editor, *British Medical Journal* (2nd April 1982) File PP/DAL/B/4/2/1, Wellcome Library, London.

Dally, A. Letter to the Editor, *The Lancet* (12th January 1982), File PP/DAL/B/4/1/1/1, Wellcome Library, London.

Dally, A. Letter to I. Munro (22nd January 1982), File PP/DAL/B/4/1/1/1 (File 1 of 2), Wellcome Library, London.

Dally, A. Letter to S. Openshaw (11th February 1982), File PP/DAL/B/1/1/1, Wellcome Library, London.

Dally, A. Letter to T. Hillier (20th September 1982), File PP/DAL/B/4/1/1/1 (File 1 of 2), Wellcome Library, London.

Dally, A. Letter to A. Trebach (3rd December 1982), File PP/DAL/B/4/1/1/1 (File 1 of 2), Wellcome Library, London.

Dally, A. Letter to Dr D. D. P. Rai (17th January 1983), File PP/DAL/B/4/1/1/1 (File 1 of 2), Wellcome Library, London.

Dally, A. Letter to S. Perrins (7th March 1983), File PP/DAL/B/4/1/1/1 (File 1 of 2), Wellcome Library, London.

Dally, A. Letter to the Editor, *GP Magazine* (8th April 1983), File PP/DAL/B/4/1/1/1 (File 1 of 2), Wellcome Library, London.

Dally, A. Letter to *The Guardian* newspaper (8th April 1983), File PP/DAL/B/4/1/1/1 (File 1 of 2), Wellcome Library, London.

Dally, A. Letter to M. Bishop (5th July 1983), File PP/DAL/B/4/1/1/1 (File 1 of 2), Wellcome Library, London.

Dally, A. Letter to N. P. Da Sylva (27th February 1984), File PP/DAL/B/4/1/1/1, Wellcome Library, London.

Dally, A. Letter to M. Thatcher (1st June 1984), File PP/DAL/B/4/1/1/1 (File 2 of 2), Wellcome Library, London.

Dally, A. Letter to Dr A. P. Gray (21st November 1986), File PP/DAL/B/4/1/1/1 (File 2 of 2), Wellcome Library, London.

Davies, M. Memo to K. Jarvie, 'Drug Misuse Clinical Guidelines – Leak of Working Papers' (1st July 1998), File 16/DRU 323/22 Vol 1, Department of Health Archive, Nelson, Lancashire.

Department of Health and Social Security, 'Drug Dependence: Clinical Conference', [Minutes of meeting] (28th November 1968), Private archive.

Department of Health and Social Security, 'Heroin Dependence: Clinical Conference', [Minutes of meeting] (18th September 1969), Private archive.

Department of Health and Social Security, 'Heroin Dependence: Clinical Conference, 1st December 1969', [Minutes of meeting] (January 1970), Private archive.

Department of Health and Social Security, Memos written between 25th May 1976 and 7th June 1976, File D/A242/12B, Department of Health Archive, Nelson, Lancashire.

Department of Health and Social Security, 'Press release' (26th September 1977), D2/A242/12 Vol G, Department of Health Archive, Nelson, Lancashire.

Department of Health, 'Clinical Guidelines Working Group Note of meeting held on Friday 7th November 1997 at Waterbridge House' (Undated), Private archive.

Department of Health, 'Meeting to discuss Misuse of Drugs regulations and private practitioners' (June 1997), Private archive.

Department of Health, CGWG (97) 26, 'Private practice and the prescribing of controlled drugs' (June 1997), Private archive.

Department of Health, 'Paper CGWG (97) 42 Recommendations to accompany clinical guidelines' (October 1997), Private archive.

Department of Health, Clinical Guidelines Working Group, 'Draft guidelines' (3rd July 1998), [draft B], Private archive.

Department of Health, Draft sent to Clinical Guidelines Working Group on 3rd July 1998 (undated), Private archive.

Department of Health, Draft sent to Clinical Guidelines Working Group on 5th March 1998 (undated), Private archive.

Ford, C. Facsmile to D. Raistrick 'Re DoH Injecting Subgroup' (1st June 1997), Private archive.

Ford, C. Letter to J. Strang, C. Gerada, M. Farrell and M. Davies 'Re: Clinical Guidelines Working Group – Notes of the meeting of 16th March' (24th May 1998), Private archive.

Ford, C. Letter to M. Davies (18th July 1998), Private archive.

Fowler, N. Letter to L. Britton (30th November 1983), File DAC 14, Department of Health Archive, Nelson, Lancashire.

General Medical Council, Professional Conduct Committee, Day One (5th July 1983). Case of Dally, Ann Gwendolen, T A Reed & Co. [transcript], General Medical Council Archive, London. Co. [transcript], General Medical Council Archive, London.

General Medical Council, Professional Conduct Committee, Day Two (6th July 1983), Case of Dally, Ann Gwendolen, T A Reed & Co. [transcript], General Medical Council Archive, London.

General Medical Council, Professional Conduct Committee, Day One (6th March 1984) Case of Tarnesby, Herman Peter, T A Reed & Co. [transcript], General Medical Council Archive, London.

General Medical Council, Professional Conduct Committee, Day Two (7th March 1984), Case of Tarnesby, Herman Peter, T A Reed & Co. [transcript], General Medical Council Archive, London.

General Medical Council, Professional Conduct Committee, Day Three (8th March 1984), Case of Tarnesby, Herman Peter, T A Reed & Co. [transcript], General Medical Council Archive, London.

General Medical Council, Professional Conduct Committee, Day Four (9th March 1984) Case of Tarnesby, Herman Peter, T A Reed & Co. [transcript], General Medical Council Archive, London.

General Medical Council, Professional Conduct Committee, Day Five (10th March 1984) Case of Tarnesby, Herman Peter, T A Reed & Co. [transcript], General Medical Council Archive London.

General Medical Council, Professional Conduct Committee, Day One (9th December 1986), Case of Dally, Ann Gwendolen, T A Reed & Co. [transcript], General Medical Council Archive, London.

General Medical Council, Professional Conduct Committee, Day Three (11th December 1986). Case of Dally, Ann Gwendolen, T A Reed & Co. [transcript], General Medical Council Archive, London.

General Medical Council, Professional Conduct Committee (4th July 1988) Case of Dally, Ann Gwendolen (Resumed case), T A Reed & Co. [transcript], General Medical Council Archive, London.

Harbridge, E. Letter to A. Dally (14th July 1983), File PP/DAL/B/4/1/1/1 (File 1 of 2), Wellcome Library, London.

Home Office, TRWG Mins 23, Minutes of the 23rd meeting of the Treatment and Rehabilitation Working Group (27th November 1978), File D2/A242/12 Vol G, Department of Health Archive, Nelson, Lancashire.

Home Office, TRWG (82) 24 (7th May 1982), File DAC 7, Department of Health Archive, Nelson, Lancashire.

Home Office Drugs Inspectorate, 'Irresponsible prescribing enquiries: investigation, preparation and presentation of evidence' (1983), Private archive.

Home Office, 'Report of a Tribunal set up under the Misuse of Drugs Act 1971 to enquire into the conduct of Dr John Adrian Garfoot MB BS MRCS LRCP, 1994', Private archive of Dr Garfoot.

Home Office, MWG (84) 33, 'Power to restrict licences to prescribe under Misuse of Drugs Act 1971' (1984), Private archive.

Home Office, TRWG Mins 22, Advisory Council on the Misuse of Drugs, 'Treatment and Rehabilitation Working Group', Meeting held on 2nd

October 1978, File D/A242/12, Department of Health Archive, Nelson, Lancashire.

Jarman, B. Letter to A. Dally (7th April 1982), File PP/DAL/B/4/2/1, Wellcome Library, London.

Lee, P. A. DHSS minute (25th May 1976), File D/A242/12B, Department of Health Archive, Nelson, Lancashire.

Macfarlane, A. D. CGWG (97) 3, 'Pharmaceutical diversion and the prescribing dimension. note by the Home Office Drugs Inspectorate' (3rd March 1997), Private archive.

Macfarlane, A. D. CGWG (97) 17, 'Clinical Guidelines Working Group. Note by the Home Office on licensing', Home Office (8th May 1997), Private archive.

Mellor, D. Letter to R. Whitney (10th December 1985), File 16/DAC 28/2, Department of Health Archive, Nelson, Lancashire.

Munro, I. Letter to A. Dally (7th January 1982), File PP/DAL/B/1/1/1, Wellcome Library, London.

Munro, I. Letter to A. Dally (20th January 1982), File PP/DAL/B/1/1/1, Wellcome Library, London.

Munro, I. Letter to A. Dally (date missing, 1983), File PP/DAL/B/4/1/1/1 (File 1 of 2), Wellcome Library, London.

Munro, I. Letter to J. Sharp (20th March 1987), File PP/DAL/B/4/1/3/2, Wellcome Library, London.

Patten, J. Letter to D. Mellor (15th May 1985), File 16/DAC 28/2, Department of Health Archive, Nelson, Lancashire.

Patterson, D. Letter to M. Davies (8th July 1998), File 16 DRU 323/12 Vol 1, Department of Health Archive, Nelson, Lancashire.

Privy Council, Appeal No.7 of 1987, *Ann Gwendolen Dally v. The General Medical Council, Judgment of the Lords of the Judicial Committee of the Privy Council* (Delivered the 14th September 1987), General Medical Council Archive, London.

Rai, D. D. P. Letter to A. Dally (30th January 1983), File PP/DAL/B/4/1/1/1 (File 1 of 2), Wellcome Library, London.

Reed, J. L. "Meeting of Doctors Working in London Drug Dependency Treatment Centres, November 25th, 1969 at St Bartholomew's Hospital" [Minutes of meeting], Private Archive.

Samways, D. Letter to A. Dally (9th August 1982), File PP/DAL/B/4/1/1/1 (File 1 of 2), Wellcome Library, London.

Sippert, A. DHSS Minute, 17.6.1976. File D/A242/12B, Department of Health Archive, Nelson, Lancashire.

Spear, H. B. Letter to A. Dally (1st February 1982), File PP/DAL/B/4/2/1, Wellcome Library, London.

Spear, H. B. Letter to Mr Emery (17th May 1987), File PP/DAL/B/4/1/3/2, Wellcome Library, London.

Strang, J. 'Draft outline for conclusions and recommendations about the prescribing of tablets and ampoules of methadone' (undated; attached to Strang, J. Letter to J. Polkinghorne, 29th November 1995), File 16/TFD 45 Vol 1, Department of Health Archive, Nelson, Lancashire.

Strang, J. Letter to J. Polkinghorne (29th November 1995), File 16/TFD-45 Vol 1, Department of Health Archive, Nelson, Lancashire.

Thatcher, M. Letter to A. Dally (11th August 1983), File PP/DAL/B/5/1/1, Wellcome Library, London.

Thatcher, M. Letter to A. Dally (7th June 1984), File PP/DAL/B/4/1/1/1 (File 2 of 2), Wellcome Library, London.

Thorley, A. TRWG (82) 15, Memorandum to David Hardwick (22nd March, 1982), File DAC 28, Department of Health Archive, Nelson, Lancashire.

Thorley, A. CGWG (97) 5, 'Approaches to the definition of a specialist' (26th Feb 1997), Private archive.

Thorley, A. CGWG (97) 17, 'Clinical Guidelines Working Group (attached to CGWG (97)17 'Note by the Home Office on Licensing'), Home Office Licensing: The Options for Encouraging Good Clinical Practice') (May 1997), File 16 TFD-46 Vol 5, Department of Health Archive, Nelson, Lancashire.

Turner, D. [SCODA], TRWG (2)/20 'Treatment and Rehabilitation Working Group' (16th November 1978), File D/A242/12, Department of Health Archive, Nelson, Lancashire.

Webb, L. Letter to E. Shore, Deputy Chief Medical Officer (12th October 1984), File DAC 28, Department of Health Archive, Nelson, Lancashire.

Witney, R. Letter to D. Mellor (22nd November 1985), File 16/DAC 28/2, Department of Health Archive, Nelson, Lancashire.

Oral history interviews

Appelbe, G. E. Interview by Sarah Mars (2004).
Banks, A. Interview by Sarah Mars (2001).
Beckett, H. D. Interview by Sarah Mars (2001).
Bewley, T. Interview by Sarah Mars (2001).
Brewer, C. Interview by Sarah Mars (2000).
Brewer, C. Interview by Sarah Mars (2003).
Dally, A. Interview by Sarah Mars (2002).
Doctor 005. Interview by Sarah Mars (2000).
Doctor 010. Interview by Sarah Mars (2000).
Doctor 011. Interview by Sarah Mars (2001).
Fazey, C. Interview by Sarah Mars (2002).
Ford, C. Interview by Sarah Mars (2002).
Garfoot, A. Interview by Sarah Mars (2000).
Gruer, L. Interview by Sarah Mars (2003).
Hampton, H. Interview by Sarah Mars (2002).
Hay, L. Interview by Sarah Mars (2003).
Hayward, S. L. Interview by Sarah Mars (2003).
Home Office Inspector, Interview by Sarah Mars (2002).
Johnson, M. Interview by Sarah Mars (2000).
Johnson, M. Interview by Sarah Mars (2001).
Johnson, M. Interview by Sarah Mars (2010).
Lutener, S. Interview by Sarah Mars (2003).
Macfarlane, A. Interview by Sarah Mars (2002).
Mack, J. Interview by Sarah Mars (2003)
Mitcheson, M. Interview by Sarah Mars (2003).
Patient 001. Interview by Sarah Mars (2001).

Patient 002. Interview by Sarah Mars (2002).
Sathananthan, K. Interview by Sarah Mars (2001).
Senior Civil Servant. DHSS, Interview by Sarah Mars (2001).
Senior Civil Servant. Home Office, Interview by Sarah Mars (2002).
Sharpe, M. Interview by Sarah Mars (2001).
Spurgeon, P. Interview by Sarah Mars (2004).
Strang, J. Interview by Sarah Mars (2002).
Strang, J. Interview by Sarah Mars (2010).
Thorley, A. Interview by Sarah Mars (2002).
Tilley, N. Interview by Sarah Mars (2002).
Turner, D. [SCODA], Interview by Sarah Mars (2002).
UK Charity Worker. Interview by Sarah Mars (2002).
Waller, T. Interview by Sarah Mars (2001).
Willis, J. Interview by Sarah Mars (2003).
Working Group member, Interview by Sarah Mars (2002).

Correspondence with the author

ACMD member, Personal communication (1997).
Home Office, Personal communication (2002).
Johnson, M. Personal communication (2011)
Mott, J. Personal Communication (2005).
Senior Compliance Office [Drugs Licensing & Compliance Unit], Personal communication (2010).
Strang, J. Personal communication (2000).
Turner, D. [SCODA], Personal communication (2003).

Grey literature/reports

Acheson, E. D. *'Guidelines of Good Clinical Practice in the Treatment of Drug Dependence'* [letter to all doctors accompanying the *Guidelines*] (29th October 1984).
Advisory Council on the Misuse of Drugs, Treatment and Rehabilitation Working Group, *First Interim Report* (London: DHSS, 1977).
Advisory Council on the Misuse of Drugs, *Treatment and Rehabilitation*, Department of Health and Social Security (London: HMSO, 1982).
Advisory Council on the Misuse of Drugs, *Prevention* (London: HMSO, 1984).
Advisory Council on the Misuse of Drugs, *AIDS and Drug Misuse* (London: Department of Health and Social Security, 1988).
Advisory Council on the Misuse of Drugs, *AIDS and Drug Misuse Part 2* (London: HMSO, 1989).
Advisory Council on the Misuse of Drugs, *AIDS and Drug Misuse Update*, Department of Health (London: HMSO, 1993).
Anderson, S. 'Health professionals and health care systems: the role of the state in the development of community pharmacy in Great Britain 1900 to 1990',

National Health Policies in Context Workshop (Bergen, Norway, 27–28th March 2003).

Baker, R. 'British and American Conceptions of Medical Ethics, 1847–1947', Anglo-American Medical Relations: Historical Insights Conference, 19th–21st June 2003, The Wellcome Building, London, UK.

British Medical Association, General Medical Services Committee, *Core Services: Taking The Initiative* (London: British Medical Association, 1996).

Committee of Inquiry into the Regulation of the Medical Profession, *Report of the Committee of Inquiry into the Regulation of the Medical Profession* [Merrison Report] (London: HMSO, 1975).

Departmental Committee on Morphine and Heroin Addiction, Report [Rolleston Report] (London: HMSO, 1926).

Department of Health and Social Security, *Better Services for the Mentally Ill* (London: HMSO, 1975).

Department of Health and Social Security, Department of Education and Science, Home Office and Manpower Services Commission, *Misuse of Drugs with Special Reference to the Treatment and Rehabilitation of Misusers of Hard Drugs: Government Response to the Fourth Report from the Social Services Committee Session 1984–85* (London: HMSO, 1985).

Department of Health, Scottish Office Home and Health Department and Welsh Office, *Drug Misuse and Dependence. Guidelines on Clinical Management* (London: HMSO, 1991).

General Medical Council, *Minutes of the General Medical Council and Committees for the Year 1979 with Reports of the Committees, etc.* CXVI (London: General Medical Council, 1979).

General Medical Council, *Minutes of the General Medical Council and Committees for the Year 1991 with Reports of the Committees, etc.* CXXVIII (London: General Medical Council, 1991).

HM Government, *Safer Management of Controlled Drugs. The Government's Response to the Fourth Report of the Shipman Inquiry* (London: HM Government, 2004) pp. 16–17. http://www.dh.gov.uk/en/Publicationsandstatistics/Publications/PublicationsPolicyAndGuidance/DH_4097904

Home Office Drugs Branch Inspectorate, *Annual Report 1985* (London: Home Office, 1986) Ref. 52186, DrugScope Library, London.

Home Office Drugs Branch Inspectorate, *Annual Report 1986* (London: Home Office, 1987), Ref. 49910, DrugScope Library, London.

Home Office Drugs Branch Inspectorate, *Annual Report 1987* (London: Home Office, 1988), Ref. 52150, DrugScope Library, London.

House of Commons Social Services Committee, *Misuse of Drugs with Special Reference to the Treatment and Rehabilitation of Misusers of Hard Drugs. Fourth Report of the Social Services Committee, Session 1984–85* (London: HMSO, 1985).

Interdepartmental Committee on Drug Addiction, *Drug Addiction. The Second Report of the Interdepartmental Committee* [second Brain Report], Ministry of Health, Scottish Home and Health Department (London: HMSO, 1965).

Labour Party, *Drugs: The Need for Action* (London: Labour Party, 1994).

Lord President of the Council and Leader of the House of Commons, Secretary of State for the Home Department, Secretary of State, et al., *Tackling Drugs Together. A Strategy for England 1995–1998* (London: HMSO, 1995).

Lord Privy Council, *Tackling Drug Misuse: A Summary of the Government's Strategy* (London: HMSO, 1985).

Macfarlane, A. 'Changes to the misuse of drugs legislation of controlled drugs prescribe in the treatment of addiction', [Consultation document] (London: Action Against Drugs Unit, Home Office, 17th March 2000).

Mars S. and Berridge, V. 'Social Science Small Grant End of Project Report Ref No: SGS/00742/G, The impact of drug user patient groups ('user groups') on UK drug treatment policy since the 1970s.' [unpublished] (2002).

Mars, S. 'Public versus private treatment for addiction: Britain in the 1980s', *Health Between the Private and the Public – Shifting Approaches*. European Association for the History of Medicine and Health Annual Conference (Oslo, Norway: 3rd–7th September 2003).

Medical Working Group on Drug Dependence, *Guidelines of Good Clinical Practice in the Treatment of Drug Misuse* (London: Department of Health and Social Security, 1984).

National Treatment Agency for Substance Misuse, Prescribing Services for Drug Misuse (London: NTASM, 2003).

Spurgeon, P. 'Crack and other drugs – myth or menace?' *Looking Ahead 1989 Conference Report* (Swindon: Swindon Crime Concern, 1989).

Task Force to Review Services for Drug Misusers, *Report of an Independent Review of Drug Treatment Services in England*, Department of Health (London: The Stationery Office, 1996).

The Shipman Inquiry, *Fourth Report – The Regulation of Controlled Drugs in the Community* (London: HMSO, 2004).

Uchtenhagen, A. Gutzwiller, F. and Dobler-Mikola, A. *Programme for a Medical Prescription of Narcotics: Final Report of the Research Representatives. Summary of the Synthesis Report* (Zurich: University of Zurich, 1997).

UK Health Departments, *Drug Misuse and Dependence. Guidelines on Clinical Management* (London: The Stationery Office, 1999).

Published sources

Allsop, J. 'Regulation and the medical profession', in J. Allsop and M. Saks (eds.), *Regulating the Health Professions* (London; Thousand Oaks, CA; and New Delhi: Sage Publications, 2002), pp. 79–93.

Anderson, S. and Berridge, V. 'Opium in 20th-century Britain: pharmacists, regulation and the people', *Addiction*, 95 (2000), 23–36.

Anonymous, 'Doctor death', *The Listener* (29th July 1982), 22.

Anonymous, 'What's happening with heroin?', *Druglink Information Letter*, Institute for the Study of Drug Dependence, 17 (1982), 1–5.

Anonymous, 'Heroin on the NHS', *New Society*, 1269 (1987), 3.

Anonymous, 'Journal Interview 27: conversation with Philip Connell', *British Journal of Addiction*, 85 (1990), 13–23.

Anonymous, 'Journal Interview 36: conversation with Thomas Bewley', *Addiction*, 90 (1995), 883–892.

Armstrong, D. *Political Anatomy of the Body. Medical Knowledge in the Twentieth Century* (Cambridge: Cambridge University Press, 1983).

Ashton, M. 'Controlling addiction: the role of the Clinics', *Druglink Information Letter*, Institute for the Study of Drug Dependence, 13, 1 (1980), 1–6.

Ashton, M. 'Doctors at war', *Druglink*, 1, 1 (1986), 13–15.

Ashton, M. 'Doctors at war', *Druglink*, 1, 2 (1986), 14–16.

Aylett, P. 'Prescription of controlled drugs to addicts' [letter], *British Medical Journal*, 287 (1983), 127.

Baggott, R. 'Regulatory politics, health professionals, and the public interest', in J. Allsop and M. Saks (eds.), *Regulating the Health Professions* (London; Thousand Oaks, CA; and New Delhi: Sage Publications, 2002), pp. 31–47.

Ball, S. A. 'Personality traits, disorders, and substance abuse', in R. M. Stelmack (ed.), *On The Psychobiology of Personality: Essays in Honor of Marvin Zuckerman* (Oxford: Elsevier, 2004), pp. 203–222.

Banks, A. and Waller, T. A. N. *Drug Addiction and Polydrug Use: The Role of the General Practitioner* (London: Institute for the Study of Drug Dependence, 1983).

Bean, P. 'Policing the medical profession: the use of Tribunals', in D. K. Whynes and P. T. Bean (eds.), *Policing and Prescribing. The British System of Drug Control* (Basingstoke: Macmillan, 1991), pp. 60–70.

Beaumont, B., Carnworth, T., Clee, W. *et al.* 'Licensing doctors counters the National Strategy', *Druglink*, 15, 6 (2001), 25.

Beckett, H. D. 'Heroin, the gentle drug', *New Society*, 49, 877 (1979), 181–182.

Beckett, H. D. 'Prescription of controlled drugs to addicts' [letter], *British Medical Journal*, 287 (1983), 127.

Bennett, S., McPake, B. and Mills, A. *Private Providers in Developing Countries: Serving the Public Interest?* (Atlantic Highlands, NJ: Zed Books, 1997).

Bennett, T. and Wright, R. 'Opioid users' attitudes towards and use of NHS clinics, general practitioners and private doctors', *British Journal of Addiction*, 81 (1986), 757–763.

Berridge, V. 'War conditions and narcotics control: the passing of Defence of the Realm Act Regulation 40B', *Journal of Social Policy*, 19 (1978), 285–304.

Berridge, V. 'Drugs and social policy: the establishment of drug control in Britain, 1900–1930', *British Journal of Addiction*, 79 (1984), 17–29.

Berridge, V. 'Historical issues', in S. MacGregor (ed.), *Drugs and British Society. Responses to a Social Problem in the 1980s* (London and New York: Routledge, 1989), pp. 20–35.

Berridge, V. 'The 1940's and 1950's: the rapprochement of psychology and bio-chemistry, The Society for the Study of Addiction 1884–1988', *British Journal of Addiction*, Special Issue, 85, 8 (1990), 1037–1052.

Berridge, V. 'AIDS and British drug policy: continuity or change?', in V. Berridge and P. Strong (eds.), *AIDS and Contemporary History* (Cambridge: Cambridge University Press, 1993), pp. 135–156.

Berridge, V. *AIDS in the UK. The Making of Policy, 1981–1994* (Oxford: Oxford University Press, 1996).

Berridge, V. 'Doctors and the state: the changing role of medical expertise in policy-making', *Contemporary British History*, 11 (1997), 66–85.

Berridge, V. 'AIDS and the rise of the patient? Activist organisation and HIV/AIDS in the UK in the 1980s and 1990s', *Medizin Gesellschaft und Geschichte*, 21 (2002), 109–123.

Berridge, V. 'Issue network versus producer network? ASH, the Tobacco Products Research Trust and UK smoking policy', in V. Berridge (ed.), *Making Health Policy: Networks in Research and Policy After 1945* (Amsterdam: Rodopi, 2005), pp. 101–124.

Berridge, V. and Edwards, G. *Opium and the People. Opiate Use in Nineteenth-Century England* (first published 1981) (London: Allen Lane; this edition New Haven CT and London: Yale University Press, 1987).

Berridge, V. and Mars, S. 'Glossary of the history of addiction', *Journal of Epidemiology and Community Health*, 58 (2004), 747–750.

Berridge, V. and Webster, C. 'The crisis of welfare, 1974–1990s', in C. Webster (ed.), *Caring for Health: History and Diversity* (Buckingham: Open University Press, 1993), pp. 127–149.

Berridge, V. *Opium and the People. Opiate Use and Drug Control Policy in Nineteenth and Early Twentieth Century England* (first published 1981, London: Allen Lane; this edition London: Free Association Books, 1999).

Berridge, V., Webster, C. and Walt, G. 'Mobilisation for total welfare 1948–1974', in C. Webster (ed.), *Caring for Health: History and Diversity* (Buckingham: Open University Press, 1993), pp. 107–126.

Bewley, T. 'Drug dependence in the USA', *Bulletin on Narcotics*, XXI, 2 (1969), 13–29.

Bewley, T. H. 'The illicit drug scene', *British Medical Journal*, 2 (1975), 318–320.

Bewley, T. H. 'Prescribing psychoactive drugs to addicts', *British Medical Journal*, 281 (1980), 497–498.

Bewley, T. H. and Ghodse, A. H. 'Unacceptable face of private practice: prescription of controlled drugs to addicts', *British Medical Journal*, 286 (1983), 1876–1877.

Bewley, T. H. and Fleminger, R. S. 'Staff/patient problems in drug dependence treatment clinics', *Journal of Psychosomatic Research*, 14 (1970), 303–306.

Brahams, D. 'No right of appeal against GMC finding of serious professional misconduct without suspension or erasure', *The Lancet*, 2 (1983), 979–981.

Brahams, D. '"Serious professional misconduct" in relation to private treatment of drug dependence', *The Lancet*, 1 (1987), 340–341.

British Medical Association, *The Misuse of Drugs* (Amsterdam: Harwood Academic Publishers, 1997).

British Medical Journal, 'Doctors for drug addicts', *British Medical Journal*, 286 (1983), 1844.

Burr, A. 'The Piccadilly drug scene', *British Journal of Addiction*, 78, 1 (1983), 5–19.

Butler, S. 'The making of the methadone protocol: the Irish system?', *Drugs: Education, Prevention and Policy*, 9, 4 (2002), 311–324.

Cawley, R. 'Obituary: Dr P H Connell', *The Independent*, Monday 10th August 1998.

Connell, P. H. 'Drug dependence in Great Britain: a challenge to the practice of medicine', in H. Steinberg (ed.), *Scientific Basis of Drug Dependence*, Coordinating Committee for Symposia on Drug Action (London: J&A Churchill, 1969), pp. 291–299.

Connell, P. H. '1985 Dent Lecture: "I need heroin". Thirty years' experience of drug dependence and of the medical challenges at local, national, international and political level. What next?', *British Journal of Addiction*, 81 (1986), 461–472.

Connell, P. H. 'Treatment of drug-dependent patients, 1968–1969', *British Journal of Addiction*, 86 (1991), 913–915.

Connell, P. H. and Mitcheson, M. 'Necessary safeguards when prescribing opioid drugs to addicts: experience of drug dependence clinics in London', *British Medical Journal*, 288 (1984), 767–769.

Cooter, R. 'The ethical body', in R. Cooter and J. Pickstone (eds.), *Medicine in the Twentieth Century* (Amsterdam: Harwood Academic Publishers, 2000), pp. 457–468.

Cox, N. 'National British archives: public records', in B. Brivati, J. Buxton and A. Seldon (eds.), *The Contemporary British History Handbook* (Manchester and New York: Manchester University Press, 1996), pp. 253–271.

Curson, D. 'Private treatment of alcohol and drug problems in Britain', *British Journal of Addiction*, 86 (1991), 9–11.

Dally, A. 'Personal view', *British Medical Journal*, 283 (1981), 857.

Dally, A. 'Drug clinics today' [letter], *The Lancet*, 8328 (1983), 826.

Dally, P. 'Unacceptable face of private practice: prescription of controlled drugs to addicts' [letter], *British Medical Journal*, 287 (1983), 500.

Dally, A. *A Doctor's Story* (London: Macmillan, 1990).

Davies, J. B. *The Myth of Addiction* (Amsterdam: Harwood Academic Publishers, 1997).

Dorn, N., Baker, O. and Seddon, T. *Paying for Heroin: Estimating The Financial Cost of Acquisitive Crime Committed by Dependent Heroin Users in England and Wales* (London: Institute for the Study of Drug Dependence, 1994).

Dorn, N. and South, N. (eds.) *A Land Fit for Heroin? Drug Policies, Prevention and Practice* (Basingstoke: Macmillan, 1987).

Douglas, M. *Cultural Bias*, Occasional Paper No. 35 (London: Royal Anthropological Institute of Great Britain and Northern Ireland, 1978).

Douglas, M. *Cultural Bias*, Occasional Paper No. 35 (London: Royal Anthropological Institute, first published 1978, second impression 1979).

Douglas, M. and Wildavsky, A. *Risk and Culture. An Essay on the Selection of Technological and Environmental Dangers* (Berkeley; Los Angeles, CA; and London: University of California Press, 1982).

Driver, F. 'Bodies in space. Foucault's account of disciplinary power', in C. Jones and R. Porter (eds.), *Power, Medicine and the Body* (London and New York: Routledge, 1998, first published 1994), pp. 113–131.

Drury, M. 'The general practitioner and professional organisations', in I. London, J. Horder and C. Webster (eds.), *General Practice Under the National Health Service, 1948–1997* (London: Clarendon Press, 1998), pp. 205–223.

Duke, K. *Drugs, Prisons and Policy-Making* (London: Macmillan, 2003).

Duval, M. K. and Den Boer, J. 'Consumer health education', in A. Levin (ed.), *Regulating Health Care. The Struggle for Control Proceedings of the Academy of Political Science, Vol 33, No. 4* (New York: Academy of Political Science, 1980), pp. 168–181.

Edwards, G. and Gross, M. 'Alcohol dependence: provisional description of a clinical syndrome', *British Medical Journal*, 8 (1976), 1058–1061.

Foucault, M. *Discipline and Punish. The Birth of the Prison*, translated by A. Sheridan (Edition Gallimard, first published 1975, this edition Harmondsworth, England: Penguin Books, 1991).

Foucault, M. *The Birth of the Clinic. An Archaeology of Medical Perception*, translated by A. M. Sheridan Smith (Presses Universitaires de France, first published 1963, this edition New York: Vintage Books, 1994).

Friedson, E. 'The centrality of professionalism to health care', in E. Friedson (ed.), *Professionalism Reborn. Theory, Prophesy and Policy* (Cambridge: Polity Press with Blackwell Publishers, 1994), pp. 184–198.

General Medical Council, *Professional Conduct and Discipline: Fitness to Practice*, edition 61 (London: General Medical Council, 1985).

Ghodse, A. H. 'Drug problems dealt with by 62 London casualty departments', *British Journal of Preventive and Social Medicine*, 30 (1976), 251–256.

Ghodse, A. H. 'Treatment of drug addiction in London', *The Lancet*, 1 (1983), 636–639.

Ghodse, H., Oyefeso, A. and Kilpatrick, B. 'Mortality of drug addicts in the UK, 1967–1993', *International Journal of Epidemiology*, 27 (1998), 473–478.

Glanz, A. 'The fall and rise of the general practitioner', in J. Strang and M. Gossop (eds.), *Heroin Addiction and Drug Policy: The British System* (Oxford, New York, and Tokyo: Oxford University Press, 1994), pp. 151–166.

Gluckman, M. *Custom and Conflict in Africa* (Oxford: Basil Blackwell, first published 1956, this edition, 1982).

Ham, C. *Health Policy in Britain: The Politics and Organisation of the National Health Service* (first published 1982, London: Macmillan; fourth edition Houndsmill: Macmillan, 1999).

Hansard, House of Lords (14th January 1971), Vol 314, col. 245.

Hansard, House of Lords (14th January 1971), Vol 314, col. 248.

Harrison, S. and Ahmad, W. I. U. 'Medical autonomy and the UK state 1975–2025', *Sociology*, 34 (2000), 129–146.

Hartnoll, R. and Lewis, R. 'Unacceptable face of private practice: prescription of controlled drugs to addicts', [letter], *British Medical Journal*, 287 (1983), 500.

Hartnoll, R., Mitcheson, M. C., Battersby, A. *et al.* 'Evaluation of heroin maintenance in controlled trial', *Archives of general psychiatry*, 37 (1980), 877–884.

Hawes, A. J. 'Goodbye junkies. A general practitioner takes leave of his addicts', *The Lancet*, 1 (1970), 258–260.

Hawks, D. 'The dimensions of drug dependence in the United Kingdom', in G. Edwards, M. A. H. Russell, D. Hawks *et al.* (eds.), *Drugs and Drug Dependence* (Farnborough, Hampshire, England: Saxon House and Lexington, MA: Lexington Books, 1976), pp. 5–29.

Hay, C. and Richards, D. 'The tangled webs of Westminster and Whitehall: the discourse, strategy and practice of networking within the British core executive', *Public Administration*, 78 (2000), 1–28.

Heclo, H. and Wildavsky, A. *The Private Government of Public Money. Community and Policy Inside British Politics* (London: Macmillan, first editon 1974, this edition, 1981).

Hoare, J. and Moon, D. (eds.) *Drug Misuse Declared: Findings from the 2009/10 British Crime Survey England and Wales* (London: Home Office, 2010).

Honigsbaum, F. *The Division in British Medicine. A History of the Separation of General Practice from Hospital Care 1911–1968* (New York: St Martin's Press, 1979).

Honigsbaum, M. 'The addiction arguments that divide the doctors', *Hampstead and Highgate Express* (6th May 1983), 2.

Hudebine, H. 'Applying cognitive policy analysis to the drug issue: harm reduction and the reversal of the deviantization of drug users in Britain 1985–2000', *Addiction Research and Theory*, 13 (2005), 231–243.

Institute for the Study of Drug Dependence, *Drug Misuse in Britain 1996* (London: Institute for the Study of Drug Dependence, 1997).

James, M. 'Historical research methods', in K. McConway (ed.), *Studying Health and Disease* (London: Open University Press, 1994), pp. 36–48.

Jeffrey, C. G. 'Drug control in the United Kingdom', in R. V. Phillipson (ed.), *Modern Trends in Drug Dependence and Alcoholism* (London: Butterworths, 1970), pp. 60–74.

Johnson, T. 'Governmentality and the institutionalization of expertise', in T. Johnson, G. Larkin and M. Saks (eds.), *Health Professions and the State in Europe* (London: Routledge, 1995), p. 237.

Johnson, T. J. *Professions and Power* (London and Basingstoke: Macmillan Press, 1972).

Kerr, A. 'In conversation with Thomas Bewley', *Psychiatric Bulletin*, 31 (2007), 220–223.

Klein, R. *The Politics of the National Health Service* (London and New York: Longman, 1983).

Klein, R. *The New Politics of the National Health Service* (London and New York: Longman, first published 1983, third edition 1995).

Klein, R. *The New Politics of the National Health Service* (first published 1983, London and New York: Longman; fourth edition, Harlow, Essex: Prentice Hall, 2001).

Laing and Buisson, *Review of Private Healthcare and Long Term Care* (London: Laing and Buisson, 1999).

Laing, W. *Going Private. Independent Health Care in London* (London: King's Fund, 1992).

Laurance, J. and Dally, A. 'Racketeer or rescuer?', *New Society*, 1256 (1987), 18–19.

Lawton, D. and Gordon, P. *HMI* (London and New York: Routledge and Kegan Paul, 1987).

Leech, K. 'Bing Spear: appreciations', in H. B. Spear and J. Mott (eds.), *Heroin Addiction Care and Control: The 'British System' 1916–1984* (London: DrugScope, 2002), pp. ix–xi.

Lewis, R. 'Flexible hierarchies and dynamic disorder – the trading and distribution of illicit heroin in Britain and Europe, 1970–1990', in J. Strang and M. Gossop (eds.), *Heroin Addiction and Drug Policy: The British System* (Oxford, New York, and Tokyo: Oxford University Press, 1994), pp. 42–54.

Lewis, R., Hartnoll, R., Bryer, S. *et al.*, 'Scoring smack: the illicit heroin market in London 1980–1983', *British Journal of Addiction*, 80 (1985), 281–290.

Lindblom, C. E. 'The science of "muddling through"', *Public Administration Review*, 19, 2 (1959), 79–88.

MacGregor, S. (ed.) *Drugs and British Society. Responses to a Social Problem in the 1980s* (London and New York: Routledge, 1989).

MacGregor, S. 'Choices for policy and practice', in S. MacGregor (ed.), *Drugs and British Society. Responses to a Social Problem in the 1980s* (London and New York: Routledge, 1989), pp. 170–200.

MacGregor, S. 'Pragmatism or principle? Continuity and change in the British approach to treatment and control', in R. Coomber (ed.), *The Control of Drugs and Drug Users. Reason or Reaction?* (Amsterdam: Harwood Academic Publishers, 1998), pp. 131–154.

MacGregor, S. 'Policy responses to the drugs problem', in S. MacGregor (ed.), *Responding to Drug Misuse. Research and Policy Priorities in Health and Social Care* (London and New York: Routledge, 2010), pp. 1–13.

MacGregor, S., Ettorre, B., Coomber, R. *et al.*, *Drug Services in England and the Impact of the Central Funding Initiative*, Institute for the Study of Drug Dependence Research Monograph One (London: Institute for the Study of Drug Dependence, 1991).

Mars, S. 'Peer pressure and imposed consensus: the making of the 1984 *Guidelines of Good Clinical Practice in the Treatment of Drug Misuse*', in V. Berridge (ed.), *Making Health Policy: Networks in Research and Policy After 1945* (Amsterdam: Rodopi, 2005), pp. 149–182.

McElrath, K. *Heroin Use in Northern Ireland: A Qualitative Study into Heroin users' Lifestyles, Experiences andRrisk Behaviours (1997–1999)* (Northern Ireland: Department of Health, Social Services and Public Safety, 2001).

McKeganey, N. 'Shadowland: general practitioners and the treatment of opiate-abusing patients', *British Journal of Addiction*, 83 (1988), 373–386.

Merritt, J. 'Doctors who trade in misery', *Daily Mirror* (18th February 1982), 7–8.

Milner, G. 'Prescription of controlled drugs to addicts' [letter], *British Medical Journal*, 287 (1983), 127.

Mitcheson, M. 'Drug clinics in the 1970s', in J. Strang and M. Gossop (eds.), *Heroin Addiction and Drug Policy: The British System* (Oxford, New York, and Tokyo: Oxford University Press, 1994), pp. 178–191.

Mold, A. *Heroin. The Treatment of Addiction in Twentieth-century Britain* (DeKalb, IL: Northern Illinois University Press, 2008).

Mold, A. and Berridge, V. *Voluntary Action and Illegal Drugs. Health and Society in Britain Since the 1960s* (Houndsmill, Basingstoke: Palgrave Macmillan, 2010).

Moran, M. *Governing the Health Care State* (Manchester: Manchester University Press, 1999).

Moran, M. and Wood, B. *States, Regulation and the Medical Profession* (Buckingham and Bristol, PA: Open University Press, 1993).

Morris, S. 'Doctors accused on heroin advice', *The Guardian* newspaper (24th February 2004).

Moscucci, O. *The Science of Woman* (Cambridge: Cambridge University Press, 1993).

Mott, J. 'Notification and the Home Office', in J. Strang and M. Gossop (eds.), *Heroin Addiction and Drug Policy: The British System* (Oxford, New York, and Tokyo: Oxford University Press, 1994), pp. 270–291.

Musto, D. *The American Disease. Origins of Narcotic Control* (New York and Oxford: Oxford University Press, 1999, third edition, first edition 1973).

Musto, D. 'Bing Spear: appreciations', in H. B. Spear and J. Mott (eds.), *Heroin Addiction Care and Control: The 'British System' 1916–1984* (London: DrugScope, 2002), pp. viii–ix.

NHS Executive, *Regulating Private and Voluntary Healthcare. A Consultation Document* (London: Department of Health, 1999).

O'Donnell, M. 'One man's burden', *British Medical Journal*, 287 (1983), 990.

O'Donnell, M. 'One man's burden', *British Medical Journal*, 294 (1987), 451.

Oldroyd, D. 'By grid and group divided: Buckland and the English geological community in the Early Nineteenth Century (Essay Review)', *Annals of Science*, 41 (1984), 383–393.

Orford, J. 'Addiction as excessive appetite', *Addiction*, 96 (2001), 15–31.

Parker, H., Bury, C. and Egginton, R. *New Heroin Outbreaks Amongst Young People in England and Wales*, Crime Detection and Prevention Series, Paper 92 (London: Home Office Police Policy Directorate, 1998).

Pearson, G. 'Drugs at the end of the century', *British Journal of Criminology*, 39 (1999), 477–487.

Plant, M. 'The epidemiology of illicit drug-use', in S. MacGregor (ed.), *Drugs and British Society. Responses to a Social Problem in the 1980s* (London and New York: Routledge, 1989), pp. 52–63.

Porter, D. and Porter, R. *Patient's Progress: Doctors and Doctoring in Eighteenth-Century England* (Stanford, CA: Stanford University Press, 1989).

Porter, R. 'Thomas Gisborne: physicians, Christians and gentlemen', in A. Wear, J. Geyer-Kordesch and R. French (eds.), *Doctors and Ethics: The Earlier Historical Setting of Professional Ethics* (Amsterdam and Atlanta, GA: Rodopi, 1993), pp. 252–273.

Power, R. 'Drug trends since 1968', in J. Strang and M. Gossop (eds.), *Heroin Addiction and Drug Policy: The British System* (Oxford, New York, and Tokyo: Oxford University Press, 1994), pp. 29–41.

Power, R. 'Drugs and the media: prevention campaigns and television', in S. MacGregor (ed.), *Drugs and British Society. Responses to a Social Problem in the 1980s* (London and New York: Routledge, 1989), pp. 129–142.

Preston, A. and Bennett, G. 'The history of methadone and methadone prescribing', in G. Tober and J. Strang (eds.), *Methadone Matters: Evolving Community Methadone Treatment of Opiate Addiction* (London: Martin Dunitz, 2003), pp. 13–20.

Rhodes, G. *Inspectorates in British Government. Law Enforcement and Standards of Efficiency*, Royal Institute of Public Administration (London: George Allen and Unwin, 1981).

Rhodes, R. A. W. 'Policy networks a British perspective', *Journal of Theoretical Politics*, 2, 3 (1990), 293–317.

Ripley, R. B. and Franklin, G. A. *Congress, the Bureaucracy and Public Policy* (Pacific Grove, CA: Brooks/Cole, first published 1976, this edition, 1987).

Ritchie, J. 'Drug crazy Britain', *The Sun*, London (17th December 1980), 14–15.

Robertson, A. B. 'Prescription of controlled drugs to addicts' [letter], *British Medical Journal*, 287 (1983), 126.

Scammell, M. 'Television and contemporary history', in B. Brivati, J. Buxton and A. Seldon (eds.), *The Contemporary History Handbook* (Manchester and New York: Manchester University Press, 1996), pp. 408–422.

Scott, C. *Public and Private Roles in Health Care Systems* (Buckingham and Bristol, PA: Open University Press, 2001).

Seldon, A. (ed.) *The Contemporary British History Handbook* (Manchester and New York: Manchester University Press, 1996), pp. 408–422.

Seldon, A. 'Elite interviews', in B. Brivati, J. Buxton and A. Seldon (eds.), *The Contemporary British History Handbook* (Manchester and New York: Manchester University Press, 1996), pp. 353–365.

Shortell, S. M. 'Occupational prestige differences within the medical and allied health professions', *Social Science and Medicine*, 8 (1974), 1–9.

Smart, C. 'Social policy and drug addiction: a critical study of policy development', *British Journal of Addiction*, 79 (1984), 31–39.

Smart, C. 'Social policy and drug dependence: an historical case study', *Drug and Alcohol Dependence*, 16 (1985), 169–180.

Smith, R. G. *Medical Discipline. The Professional Conduct Jurisdiction of the General Medical Council, 1858–1990* (Oxford: Oxford University Press, 1994).

South, N. 'Tackling drug control in Britain: from Sir Malcolm Delevingne to the new drugs strategy', in R. Coomber (ed.), *The Control of Drugs and Drug Users. Reason or Reaction?* (Amsterdam: Harwood Academic Publishers, 1998), pp. 87–106.

Spear, B. 'The early years of the "British System" in practice', in J. Strang and M. Gossop (eds.), *Heroin Addiction and Drug Policy: The British System* (Oxford, New York, and Tokyo: Oxford University Press, 1994), pp. 3–28.

Spear, H. B. 'British experience in the management of opiate dependence', in M. M. Glatt and J. Marks (eds.), *The Dependence Phenomenon* (Lancaster: MTP Press, 1982), pp. 51–79.

Spear, H. B. (and ed. J. Mott) *Heroin Addiction Care and Control: The 'British System' 1916–1984* (London: DrugScope, 2002).

Stacey, M. *Regulating British Medicine: the General Medical Council* (Chichester: John Wiley and Sons, 1992).

Stacey, M. 'The General Medical Council and professional self-regulation', in D. Gladstone (ed.), *Regulating Doctors* (London: Institute for the Study of Civil Society, 2000), pp. 28–39.

Stanton, J. 'What shapes vaccine policy? The case of hepatitis B in the UK', *Social History of Medicine*, 7, 3 (1994), 427–446.

Stigler, G. J. 'The theory of economic regulation', *Bell Journal of Economics and Management Science*, 2, 1 (1971), 3–21.

Stimson, G. V. 'British drug policies in the: a preliminary analysis and suggestions for research', *British Journal of Addiction* 82, 5 (1987), 477–488.

Stimson, G. V. and Lart, R. 'The relationship between the state and local practice in the development of national policy on drugs between 1920 and 1990', in J. Strang and M. Gossop (eds.), *Heroin Addiction and Drug Policy: The British System* (Oxford, New York, and Tokyo: Oxford University Press, 1994), pp. 331–341.

Stimson, G. V. " 'Blair declares war": the unhealthy state of British drug policy', *International Journal of Drug Policy*, 11 (2000), 259–264.

Stimson, G. V. and Oppenheimer, E. *Heroin Addiction: Treatment and Control in Britain* (London: Tavistock, 1982).

Stockwell, T. 'Psychological and social basis of drug dependence: an analysis of drug-seeking behaviour in animals and dependence as learned behaviour', in G. Edwards and M. Lader (eds.), *The Nature of Drug Dependence* (Oxford: Oxford University Press, 1990), pp. 195–203.

Strang, J. 'Personal view', *British Medical Journal*, 283 (1981), 376.

Strang, J. 'Personal view', *British Medical Journal*, 284 (1982), 972.

Strang, J. " 'The British System": past, present and future', *International Review of Psychiatry*, 1 (1989), 109–120.

Strang, J. 'A model service: turning the generalist on to drugs', in S. MacGregor (ed.), *Drugs and British Society. Responses to a Social Problem in the 1980s* (London and New York: Routledge, 1989), pp. 143–169.

Strang, J., Bearn, J. and Gossop, M. 'Opiate detoxification under anaesthesia', *British Medical Journal*, 315 (1997), 1249–1250.

Strang, J., Griffiths, P. and Gossop, M. 'Heroin in the United Kingdom: different forms, different origins, and the relationship to different routes of administration', *Drug and Alcohol Review*, 16 (1997), 329–337.

Strang, J. and Sheridan, J. 'Methadone prescribing to opiate addicts by private doctors: comparison with NHS practice in south east England', *Addiction*, 96 (2001), 567–576.

Strang, J., Sheridan, J. and Barber, N. 'Prescribing injectable and oral methadone to opiate addicts: results from the 1995 national postal survey of community pharmacies in England and Wales', *British Medical Journal*, 313 (1996), 270–272.

Strang, J., Stimson, G. V. and Des Jarlais, D. C. 'What is AIDS doing to the drug research agenda?', *British Journal of Addiction*, 87 (1992), 343–346.

Strang, J. and Taylor, C. 'Different gender and age characteristics of the UK heroin epidemic of the 1990s compared with the 1980s: new evidence from analyses of national treatment data', *European Addiction Research*, 3 (1997), 43–44.

Strong, P. M. 'Doctors and dirty work – the case of alcoholism', *Sociology of Health and Illness*, 2 (1980), 24–47.

Stungo, E. 'Prescription of controlled drugs to addicts' [letter], *British Medical Journal*, 287 (1983), 126–127.

The Lancet, 'Drug addiction: British System failing', *The Lancet*, 1 (1982), 83–84.

The Lancet, 'Drug dependence in Britain: a critical time', *The Lancet*, 2 (1983), 493–494

The Telegraph, 'MPs to back Blunkett on prescription heroin', *The Telegraph* (19th May 2002).

Thom, B. *Dealing With Drink. Alcohol and Social Policy from Treatment to Management* (London and New York: Free Association Books, 1999).

Thompson, M., Ellis, R. and Wildavsky, A. *Cultural Theory* (Boulder, CO and Oxford: Westview Press, 1990).

Thompson, P. *The Voice of the Past. Oral History* (Oxford: Oxford University Press, first edition 1978, this edition, 1988).

Thompson, P. and Perks, R. *An Introduction to the Use of Oral History in the History of Medicine* (London: National Life Story Collection, 1993).

Thomson Directories, Ltd, *London Classified Trades and Professions, Telephone Directory 1968* (London: General Post Office, 1968).

Thorley, A. 'Longitudinal studies of drug dependence', in G. Edwards and C. Busch (eds.), *Drug Problems in Britain: A Review of Ten Years* (London: Academic Press, 1981), 117–169.

Travis, A. 'Coalition shelves plans for "abstinence based" drug strategy', *The Guardian*, 8th December 2010.

Trotter, T. (and ed. R. Porter) *An Essay Medical, Philosophical, and Chemical on Drunkenness and Its Effects on the Human Body* (London: Routledge, 1988).

Turnbull, P. J., McSweeney, T., Webster, R., Edmunds, M. and Hough, M. *Drug Treatment and Testing Orders: Final Evaluation Report*, Home Office Research Study 212 (London: Home Office Research, Development and Statistics Directorate, 2000).

Turner, B. *Medical Power and Social Knowledge* (London: Sage, first published 1987, this edition, 1995).

Turner, D. 'The development of the voluntary sector: no further need for pioneers?', in J. Strang and M. Gossop (eds.), *Heroin Addiction and Drug Policy: The British System* (Oxford, New York, and Tokyo: Oxford University Press, 1994), pp. 222–230.

Tylden, E. 'Care of the pregnant drug addict', *MIMS*, 1st June (1983), 1–4.

Vollmer, H. M. and Mills, D. L. (eds.) *Professionalization* (Englewood Cliffs, NJ: Prentice Hall, 1966).

Von Walden Laing, D. *HIV/AIDS in Sweden and the United Kingdom Policy Networks, 1982–1992* (Stockholm: Stockholm University, 2001).

Waller, T. and Banks, A. 'Drug abuse pull out supplement', *GP*, March 25 (1983), 27–45.

Weber, M. *From Max Weber: Essays in Sociology*, translated and edited by H. H. Gerth and C. Wright Mills (New York and Oxford: Oxford University Press, first published 1946, this edition, 1964).

Webster, C. *The National Health Service. A Political History* (Oxford: Oxford University Press, 1998).

Wells, B. 'Narcotics Anonymous (NA) in Britain', in J. Strang and M. Gossop (eds.), *Heroin Addiction and Drug Policy: The British System* (Oxford, New York, and Tokyo: Oxford University Press, 1994), pp. 240–247.

Whyte, W. F. *Street Corner Society. The Social Structure of an Italian Slum* (Chicago: University of Chicago Press, 1955).

Wildavsky, A. and Polisar, D. 'From individual to system blame: analysis of historical change in the law of torts', *Journal of Policy History*, 1 (1989), 129–155.

Willis, J. H. 'Unacceptable face of private practice: prescription of controlled drugs to addicts' [letter], *British Medical Journal*, 287 (1983), 500.

Wistow, G. 'The health service policy community. Professionals pre-eminent or under challenge?', in D. Marsh and R. A. W. Rhodes (eds.), *Policy Networks in British Government* (Oxford: Clarendon Press, 1992), pp. 51–74.

Working Party of the Royal College of Psychiatrists and the Royal College of Physicians, *Drugs, Dilemmas and Choices* (London: Gaskell, 2000).

Yaggy, D. and Anlyan, W. G. (eds.) *Financing Health Care: Competition Versus Regulation. The Papers and Proceedings of the Sixth Private Sector Conference March 23 and 24, 1981,* Duke University (Cambridge, MA: Ballinger Publishing, 1982).

Yates, J. *Private Eye, Heart and Hip. Surgical Consultants, the National Health Service, and Private Medicine* (Edinburgh: Churchill Livingstone, 1995).

Yates, R. 'A brief history of British drug policy, 1950–2001', *Drugs: Education, Prevention and Policy*, 9 (2002), 113–124.

World Wide Web sources

Anonymous [names deleted], Police memorandum sent to ACPO Crime Committee (1st October 1993), Document GA 22 00075, The Shipman Inquiry, http://www.theshipmaninquiry.org.uk

Delevingne, M. Letter to Chief Constables re the Dangerous Drugs Act, 1920, and the regulations made thereunder, Home Office, Whitehall (20th

August 1921), Document WM 17 00736, The Shipman Inquiry, http://www.theshipmaninquiry.org.uk

Department of Health, 'Government to improve care for drug-misusers – New Guidelines for Doctors', Press release reference 1999/0220 (12th April 1999), http://www.dh.gov.uk

Hayzelden, J. E. Letter to Chief Officer of Police, Re Home Office Circular 25/1980 Inspection of Pharmacies (12th March 1980), Document WM 17 00062, The Shipman Inquiry, http://www.theshipmaninquiry.org.uk

Home Office, *Guidance to Chemist Inspecting Officers* (Home Office, London: 2002), Document WM 17 00388, The Shipman Inquiry, http://www.theshipmaninquiry.org.uk

Scullion, J. Transcript of Day 149 (27th June 2003), The Shipman Inquiry, http://www.theshipmaninquiry.org.uk

Scullion, J., Witness Statement (10th March 2003), Document WS 35 00001, The Shipman Inquiry, http://www.theshipmaninquiry.org.uk

Spurgeon, P. G., Letter to all Chief Police Officers Re Notes for Chemist Inspecting Officers (17th March 1988), Document WM 17 00163, The Shipman Inquiry, http://www.theshipmaninquiry.org.uk

Staff and Agencies, 'Harold Shipman: a chronology', *The Guardian* (July 15, 2004), http://www.guardian.co.uk/archive

Stears, A. Witness Statement (31st August 1999), Document WS 34 00001, The Shipman Inquiry, http://www.theshipmaninquiry.org.uk

Theses

Mars, S. G. *Prescribing and Proscribing: The Public-Private Relationship in the Treatment of Drug Addiction in England, 1970–99* (University of London: PhD Thesis, 2005).

Index

Printed in the United States
By Bookmasters